Building Underground

Building Underground

The Design and Construction Handbook
for Earth-Sheltered Houses

by Herb Wade

Illustrations by Ken Baxter

 Rodale Press, Emmaus, Pa.

The house shown on the front cover, in a photograph by Thomas M. Eiben and a drawing by Ken Baxter, was designed and built by High Country Structures, Eagle, Colorado, and is made of 3½ Terra-Dome modules. The other photograph on the front cover was taken by Lon B. Simmons.

On the back cover, the top two photos were taken by Michael Reese Much and the bottom two by Mitchell T. Mandel.

Library of Congress Cataloging in Publication Data

Wade, Herb.
    Building underground.
    Bibliography: p.
    Includes index.
    1. Earth-sheltered houses — Design and construction.
I. Title
TH4819.E27W3 1983   690.8   82-18553

ISBN 0-87857-421-2 hardcover   2 4 6 8 10 9 7 5 3 1 hardcover
ISBN 0-87857-422-0 paperback   2 4 6 8 10 9 7 5 3 1 paperback

Contents

Acknowledgments vii

Introduction... viii
    How to Use This Book ix

Chapter 1   Living Underground: After the Cave 1
    Why Underground? 3
    Is Underground Living for You? 6

Chapter 2   Designing for the Site.. 9
    Underground Constraints 9
    Designs That Fit Sites 10
    What about Rocks? 25
    Laying Out the Floor Plan 27

Chapter 3   Solar Solutions for Underground Houses................. 32
    The Solar Source 32
    Passive Solar Design 37
    Direct-Gain, Indirect-Gain and Isolated-Gain Systems........... 41
    Solar Water-Heating Systems................................. 72

Chapter 4   The First 15 Feet 77
    Understanding Soil 77
    Soil: Its Changing Character 86
    The Underground Climate 87
    Comparing the Surface Climate with the
       Subsurface Climate... 93
    The Climate around a Building 94
    Designing for the Underground Environment 97

Chapter 5   Basic Construction Techniques: Concrete
              and Wood.. 100
    Concrete 100
    Reinforced Concrete 111
    Cast-in-Place Concrete... 118
    Prestressed Concrete 120
    Ferrocement Construction...................................... 131
    Construction Using Concrete Block 137
    Earth-Contact Houses 143
    Wood Construction... 147

Chapter 6   Waterproofing and Insulation.............................. 152
    Waterproofing 152
    Insulation 164

Chapter 7   Subterranean Subsystems: Wiring, Plumbing, Heating, Cooling,
            Ventilation and Lighting 174
    Electrical Systems 174
    Plumbing 179
    Heating Systems... 185
    Summertime Comfort Control 192
    Ventilation 199
    Lighting 206

Chapter 8   A Gallery of Underground Houses.. 213
    The Cooper House.. 214
    The Freidrichs House 221
    The Juenemann House 225
    The Laughlin House 230
    The Miley House... 235
    The Pearcey House... 241
    The Rogers House 246
    The Shepard House 250

Appendix 1: Estimating Heating Requirements for Underground Houses....... 255

Appendix 2: Manufacturers of Waterproofing Materials............................. 259

For Further Reading 263

Photograph Credits 274

Index 277

Acknowledgments

There is no way to thank everyone I am indebted to for assistance on this book, but I would be remiss if I did not extend a very special thanks to the following people: Carol Stoner for being patient far beyond the call of duty; Joe Carter for being a friend in adversity as well as an outstanding editor; Margaret Balitas for the organization, editing and assistance she provided so competently; Ken Baxter for the many well-executed drawings; Lester L. Boyer, Charles A. Lane and Bill Norton for their technical reviews and constructive criticism; and David Yamamoto and Joe King for their special support and help.

I extend my heartfelt thanks also to the numerous builders, owners of underground houses, materials suppliers and manufacturers who so freely contributed their time and opened their homes and businesses to me. Without them and their pioneering spirit, earth-sheltered housing would still be an idea instead of one of the fastest-growing building technologies in North America today.

Introduction

It was one of those great Rhode Island fall mornings. I was at the University of Rhode Island library, but I really didn't want to be there. In trying to put off my "heavy" work, I was indulging in one of my many methods of procrastination by looking through old magazines. One of them caught my eye with an article by Wolfgang Langewiesche called "There's a Gold Mine under Your House." Langewiesche had written about living underground; about how a house below the surface could be cool in the summer and warm in the winter; and about underground houses that were quiet, comfortable and filled with sunlight. He had even included a map of the United States showing how underground houses could be expected to perform in different parts of the country. While reading that article, I was surprised and delighted. Instead of reading the dry (but important for a budding microbiologist) pages of *Mycologia,* I spent the day in an orgy of reading about underground construction, about homes and energy, and about man and his energy environment. That was back in 1965, and Langewiesche's article had appeared in the August 1950 issue of *House Beautiful.* Although I tried, I am afraid I never quite got back into the biology groove, and still, 18 years later, I spend a lot of time in libraries *not* reading *Mycologia.*

What I am reading, though, tells about a great change in the way people are thinking about their houses. The building industry, long known for its massive resistance to change, seems almost overnight to be moving into passive solar and underground construction. Even in the tough financial environment of the early eighties, passive solar and underground houses are selling. The staid *Wall Street Journal* acknowledged the passive solar trend in its 1 September 1981 front-page article, "Passive Solar Homes Are Enjoying a Boom amid Housing Slump." Now, supermarket magazine shelves prominently display plans books for earth-sheltered and passive solar house designs. Almost every month, home-oriented magazines feature "new" ideas for energy savings with earth-integrated design.

There are also some emotional reasons for my interest in underground houses. I grew up on a tiny farm — what today would be called a homestead — in western Missouri. An important fixture there was the storm cellar which protected us against tornadoes. The storm cellar was also the place where my mother stored garden produce and canned foods. It was always pleasantly cool in the summer and warm in the winter. It was full of goodies, and it represented security from the dreaded funnel cloud. So I have always associated security with underground structures, but I never really thought about actually living underground until I read *The Man Who Sold the Moon,* by Robert A. Heinlein. Quite peripherally to the story line, Heinlein mentions that the hero lives in a house that is small and not pretentious aboveground but large and splendid below. I was so struck by the idea that I designed a house like that for my high school drafting class. I guess you might say I have some predisposition toward underground living. I often wonder if the concentration of underground houses in the Midwest may be partly the unconscious response of builders and designers who have had similar childhood experiences.

How to Use This Book

Building Underground is a compilation of the experiences of others as tempered and translated through my own store of knowledge and experience. Throughout, my goal is to provide you with information rather than instruction. It would have been a lot easier to take specific site and climate conditions and provide specific instructions for design and construction. While that approach is not without merit, it cannot provide the reader with the information needed to *make a decision.* My aim has been to provide many reasonable alternatives in design and construction practice so that you, with the help of local professionals, can choose the unique combination of features that fits your site and satisfies your needs.

If you are reading this book expecting to be taken by the hand and led painlessly through the design and construction of your underground dream house, you will be disappointed. On the other hand, if you want to design your own house or evaluate someone else's design, you need to know what has worked and what has not, and the information herein should be of real value. I believe that by knowing *why* things are done one way or another, you will be better able to make the decisions needed to get just what you want — the very best home for you and your family.

Remember that a house is a whole system and cannot be completely divided into separate modules. While I have separated this book into chapters to make it more readable, the book should be taken as a whole, just like a house. You should read it at least once in its entirety before concentrating on any particular section.

Building Underground is divided into eight chapters, which follow the same general steps you might follow in designing an underground house.

Chapter 1, "Living Underground: After the Cave," provides historical and technical information on man's involvement with underground living throughout the ages.

Chapter 2, "Designing for the Site," concerns the integration of an underground house with its site. More so than in an aboveground house, the underground structure is a part of the site. Analysis of the site and awareness of how site characteristics affect an underground house are important when designing an underground house.

Chapter 3, "Solar Solutions for Underground Houses," describes how this renewable energy source can easily be used in most underground houses and provides rules of thumb specifically for underground house designs. This chapter provides performance characteristics for each of the major passive system classifications.

Chapter 4, "The First 15 Feet," discusses the "climate" underground and the impact it has on the thermal comfort of an underground house. This chapter also discusses soil and water and their effects on an underground house.

Chapter 5, "Basic Construction Techniques: Concrete and Wood," outlines the various construction systems for the shell of an underground house. Concrete is, of course, a common construction material, and there are a variety of ways to use it, including monolithic pours; sectional pours; pretensioned, prestressed concrete; posttensioned, prestressed concrete; sprayed ferrocement shells; concrete block; and combinations of these methods. Wood is also a satisfactory underground building material if properly treated, and it is frequently used in what are called earth-contact houses, which have fully exposed roofs and usually earth berms on three walls.

Chapter 6, "Waterproofing and Insulation," leads the reader through a maze of materials for waterproofing and insulating an underground house and provides helpful information on what materials to use for certain soils and climates.

Chapter 7, "Subterranean Subsystems," details the special problems of wiring, plumbing, heating, cooling and ventilating underground houses.

Chapter 8, "A Gallery of Underground Houses," contains examples of real underground houses of different types and in different climates.

There are appendixes and a bibliography that provide information on underground houses. Appendix 1, "Estimating Heating Requirements for Underground Houses," provides a way to estimate the sizing requirements of direct-gain solar systems as well as auxiliary heating systems. Appendix 2, "Manufacturers of Waterproofing Materials," lists numerous manufacturers of waterproofing products suitable for underground houses. The comprehensive bibliography, with over 100 entries, suggests sources for further reading.

In preparing this book, I depended on my years of experience as a designer, builder and solar researcher, but it became clear very quickly that one person's experience could never be broad enough to cover the differences in construction needs and practices across the country. To provide the wide scope I felt was necessary, I traveled around the country during the two years that I spent researching and writing this book and talked with hundreds of homeowners, builders, designers and researchers to better understand how people in various geographical areas are solving the problems of underground construction. I found that there's an underground house for every taste, pocketbook and geographical area. Besides the personal pleasure of meeting a lot of people who share my own interests, I also saw a great many successes and a few failures and always learned something from each. Our conversations were always enjoyable, and I hope this book conveys a good measure of that enjoyment along with the hard information that I hope will be useful to you as you plan your home within the shelter of the earth.

CHAPTER 1

Living Underground: After the Cave

Although underground construction is often heralded as a "new wave" that is the latest thing in housing, sheltering within the earth is probably as old as civilization itself. Cavemen retired to their underground apartments after a long day of work for the same reasons modern man seeks shelter: comfort and security from the adverse forces of man and nature. In somewhat more recent times, around A.D. 200, Romans living in Tunisia built their homes underground to escape the intense North African heat. Those homes included such "modern" innovations as roof-pond cooling, daylighting shafts, and room shapes and coloring that distributed daylight throughout well-designed underground spaces. Some 1,400 years later, the Berbers, who also lived in Tunisia, built underground. They literally carved their homes from the soil instead of digging, building and backfilling as the Romans had, and the homes appear to have been comfortable, cool and dry. In Italy and France, homes have been built into steep hillsides for centuries.

In some cases, house fronts have been built over stone mines extending deep into the hillsides. Along the Loire River valley of France, the construction of the magnificent chateaux of the region was done with massive blocks of sandstone excavated there. The stonecutters and other artisans, farmers and watermen moved into the cavities that resulted from the excavations. When anyone needed more space, it turned out that expansion was not an expense; it was profit, because it was easy to excavate and sell more stone for construction. Although many of these dwellings are in the faces of cliffs, many others are in flatland with no aboveground construction at all. As the underground community grew, community halls, shops, schools and churches filled the underground spaces left from other stone mining. In Wieliczka, Poland, a thousand years of mining salt have resulted in an underground community in over 65 miles of caverns. Churches, a central railroad station for the local narrow-gauge railway, restaurants, a ballroom with a ceiling 100 feet high, and numerous residences fill the corridors of the mine. In this century, by the end of the Second World War, Germany had almost finished building an underground airport. In the 1960s, Sweden completed huge military/industrial installations that were carved into solid rock.

In China, the loess soils of the Honan, Kansu, Shensi and Shansi provinces contain the homes of over 10 million people who live below the fields that provide their food. In Australia, opal miners live comfortably below the desert surface that has a climate far from comfortable. Although the White Cliffs community in the Australian outback was originally housed in worked-out opal mines, many additional underground houses have been carved from the rock just

Photo 1–1: One of the "modern" design features in this Roman underground house in Tunisia is an atrium that provides daylighting.

Photo 1–2: Earth sheltering is a sensible response to the harsh climate of Matmata, Tunisia, where summer temperatures can reach 131°F. This 280-year-old underground house has a central pit atrium. The rooms were made by digging into the cliffside.

for the comfort of living underground in the harsh desert climate. The list of earth-sheltered structures continues with magnificent churches carved from rock in Ethiopia, homes cut into the desert spires of the Middle East, and now, thousands of contemporary underground houses in the United States.

Why Underground?

History tells us that there are three basic reasons why people have chosen to build underground instead of on ground. The first and, I suspect, most common reason was that the shell was already there. Certainly, that was the case with the cave dweller. He didn't create the cave; he just moved in. History reveals that people made their homes in mines and natural caves all over the world. The millions of square feet of underground warehouses and industrial space in Kansas City are the result of decades of stone mining, not purposeful underground construction projects. Indeed, most underground industrial complexes fit into this category. Few present-day residences, however, are recycled mines or natural caverns.

Another reason for building underground was a military one. The huge underground installations in Sweden, the miles of tunnels in France's Maginot Line, Hitler's underground airport, extensive civil defense shelters and incredible Space Age command posts that have been hollowed out of whole mountains have all been built in the name of defense, or war. Many, if not most, of the underground houses built in the fifties and sixties were in response to the perceived need for shelter from the effects of nuclear attack.

Finally, going underground was one response to the need to be comfortable in a world that was at times too hot, too cold, too windy and even too noisy. The terms ''back to the earth'' and ''Mother Earth'' have special relevance to the present-day underground dweller. By far, most of today's underground dwellings have been constructed for reasons like these.

Even though underground houses have been built for these various reasons for thousands of years, it is only in the last few years that truly large numbers of them have been built. That is because, for the first time in history, a large and diverse society has the reasons *and the means* to go underground. The reasons are still comfort and security — both physical and economic — and the means is modern technology with high-performance concrete, wood treated to last indefinitely underground, low-cost permanent waterproofing systems and computers to help with the problems of design and performance analysis. No longer does building underground make sense only in the desert or on the side of a steep hill. Earth-sheltered houses can be built almost anywhere today with straightforward, reasonably priced methods, resulting in houses capable of energy performance and comfort qualities at least as good as the best surface-built dwellings.

Photo 1–3: Beneath the loess soil of the Shansi Province, this contemporary Chinese underground house has a courtyard that was dug in flat terrain. Shown here is the entrance to one of several rooms that are off the courtyard. Access to the courtyard is by an earthen ramp.

Photo 1–4: This underground
house in the White Cliffs mining
community is a refuge from the severe climate of the Australian outback.

Advantages of Earth Sheltering

There are a number of purely logical reasons for building a house underground, reasons why earth-sheltered designs have distinct advantages over conventional aboveground designs.

Ease of Comfort Control: The isolation that several feet of earth provide makes it possible to heat and cool an underground house to the same comfort levels as a conventional house for a small fraction of the cost. The milder, more constant temperature below the surface and the complete absence of wind make the climate that the underground house experiences much less harsh than that on the surface.

Physical Security: Violent weather is not likely to damage a house that is protected by the earth. Violent people have a harder time entering or damaging a house that is covered with earth on almost all sides. Grass fires, forest fires or fires on adjoining properties are a lessened danger. Even added protection against nuclear blast and fallout is provided. A reduction in fallout radiation by a factor of 50 to 1,000 is possible in specially designed underground houses.

Economic Security: Low exterior maintenance costs, low operating costs and extraordinarily sturdy construction mean a house with continuing value and practicality. Underground houses designed with energy conservation in mind regularly have heating and cooling bills less than one-quarter those of surface dwellings of comparable cost. Also, insurance costs are often lower for underground houses, since damage due to wind and fire is not likely to be as great as for an above-grade house.

Sound Control: In noisy environments near highways, airports and industrial complexes, sound levels in the underground house are far lower than in surface residences. The opposite is true, too. If you are often the host for noisy parties or if you are a rock star and you practice a lot at home, disturbing the neighborhood peace will not become an unwanted part of your repertoire.

But, logic notwithstanding, while it's true that some people are enamored of the idea of living underground, others express horror at the prospect. Doubters point to the similar advantages of living in energy-efficient, *super-insulated* aboveground houses, and there has been much debate on the relative merits of both forms. Super-insulated house design is a somewhat more recent development in the field of energy-efficient housing, but, like underground house design, it has attracted a lot of interest. It basically involves including maximum levels of insulation (for example, R-40 walls, R-60 ceilings) and super-tight construction methods to minimize heating and cooling loads. These houses are so efficient that their heating needs are largely met by internal heat generation (people, cooking, lights), which reduces the need for solar heating. As a result, super-insulated designs tend to look more like conventional houses than solar houses. In the debate, the issue seems best resolved by following personal tastes. Recently, for example, on the letters-to-the-editor pages of the excellent *Earth Shelter Digest*

Photo 1–5: This underground warehouse space at the Great Midwest Corporation in Kansas City, Missouri, is ready to be used as storage space.

and Energy Report (now *Earth Shelter Living*), a series of letters written by William Shurcliff and Malcolm Wells appeared. Shurcliff, who is one of the foremost thinkers on residential energy conservation, extolled the virtues of super-insulation as a way to minimize operating costs (for heating and cooling) and to reduce the building's need for large expanses of south glazing, which isn't always desirable for certain sites. Wells, who is one of the pioneers in underground design and construction, was naturally an earth-shelter booster, for many of the same reasons expressed in this book. Well, both of these experts are right. The technical arguments really come out in a dead heat, and the choices that you make depend more on the sum of many considerations: your preferences, the site, the climate and the building budget, among others.

Is Underground Living for You?

With all the strong and sensible arguments for earth sheltering, there are also some arguments against becoming an underground dweller. I think the most powerful is emotional. If you just plain don't even want to consider living in an underground house (even with all the logic in the world pointing to the contrary), you should not build underground. If you have concerns or if you are not sure that you would like such a home, read on. Then, visit underground homes, talk with the owners, and read some more about living underground. Of all the families I have talked with, there were very few in which all of the members were

Photo 1-6: The interiors of modern underground houses are spacious, well lighted and comfortable — a far cry from the dark, damp places that cave dwellers of ancient times called home.

originally happy with the idea of living underground. Yet, after actually moving into their underground home, no one expressed dissatisfaction. In fact, those persons originally against the idea were often among the strongest in their praise of underground houses.

Misconceptions about underground houses are one reason why more people are not considering earth sheltering as a housing option. People associate living underground with wet basements, damp caves and dark holes in the ground. Also, many of the people who have lived in the ground have historically been of the "lower" class. They were the ones living in caves and abandoned mines.

There may also be an economic disincentive for building underground. While it is true that in today's market an underground house can be built for about the same cost as a comparable conventional house, it is also true that a "comparable conventional house" is really a custom house with energy-conservation features and comforts beyond those of the cheapest mass-produced subdivision house. A house built to the minimum possible standards will indeed have a lower initial cost than an underground house of similar size, but the differences in comfort, maintenance cost and operating expense will be great. While the first cost of an underground house will be higher than the lowest cost of an aboveground house, reduced maintenance and energy costs may well make the underground house a better buy in the long run.

Photo 1-7: Modern technology provides the means to build underground houses that are energy efficient and attractive. This passive solar earth-sheltered house is made of poured-in-place, posttensioned concrete. The facade is cedar and stucco.

Site conditions naturally play an important role in deciding for or against an earth-sheltered house. Although modern building technology allows both aboveground and underground construction at sites that would not have been considered practical a decade or two ago, there still are sites where building underground is at best silly and at worst so terribly expensive as to be impractical. Swampy ground, solid rock and places with perfect aboveground climates are poor candidates for underground construction.

One last reason for not building underground, even when you are emotionally ready and have a good site available, is the hassle that you may have to go through to get the job done. Archaic building codes, unfriendly neighbors, difficulty in finding design and construction professionals willing to take on the project, bankers who can't relate to earth sheltering, and even family and friends who pooh-pooh the whole idea — each of these can be a difficult hurdle to overcome. Put several of these hurdles together at one time, and the hassle factor can become more than some people can tolerate. Fortunately, most bankers, most building codes or zoning officials, and a dramatically increased body of builders and engineers now look very favorably upon underground housing. With the good press underground housing has enjoyed, with the many successful examples all over the country and with your good ideas and enthusiasm, the process of design, permit approval, financing and construction may still involve some hassles, but few more than for any other new house.

CHAPTER 2

Designing for the Site

Although it is poor practice to ignore site conditions when designing aboveground houses, it is a much more serious error to fail to design for the site when going underground. Water problems are often the result of poor integration with the site, as are excessive costs resulting from unexpectedly hitting solid rock, from excessive modifications to grade and from increased structural requirements. Insulation needs, solar capability and room layouts are also site related. This chapter examines these interactions between site and design so that you can begin the long road to the selection of a plan and structural design that will provide the most house for the least cost.

Underground Constraints

Robert A. Heinlein's acronym, TANSTAAFL (There Ain't No Such Thing As A Free Lunch), aptly applies to an underground structure. While we enjoy many advantages in the underground house, there are also certain limitations that may affect those advantages.

A common structural constraint is that which limits the span of ceiling sections between supporting walls or columns. For example, conventionally poured reinforced-concrete roofs usually require a support every 10 to 20 feet, according to the roof's structural design, while a posttensioned roof panel may span 20 to 30 feet without the need for columns or interior structural walls.

Site-related constraints are common. Sloping sites tend to dictate drainage patterns, wall placements and exposures. Sites with very deep soils may accommodate vaulted ceilings or domes, while sites with rock at 3 feet are better suited to flat-ceiling designs. The location of sewers and outlets for foundation drains will generally dictate floor depths, which in turn will have a great effect on the design of ceilings and walls. In general, the site will have to be considered more carefully in underground-house design than for surface houses. In particular, orientation may be dictated by the site, which, of course, greatly affects the building's ability to use solar heat.

Cost criteria have a different effect on underground-house design than they do on surface dwellings. For example, it is often less costly per square foot of aboveground house to build two stories. For most underground home sites, however, two stories are much more expensive than horizontal expansion. Also, a garage may be much cheaper as a surface building connected to the underground house than as an extension of the underground structure. The cost of the garage may, in addition to construction costs, include the cost of the

Photo 2-1: An underground house that's designed for its site not only looks appropriate to the surroundings but usually provides the most house for the least money as well.

driveway as well. A short, relatively level access road may be possible for a surface-built garage serving a house on a site that slopes away from the street, while an underground garage built on the same site may need a long, steep and expensive driveway for access.

What this all means is simple: Either find a lot to fit your dream home or design your dream home to fit your lot! If the site and your design are not very compatible, the cost will skyrocket.

Designs That Fit Sites

As has been discussed, it is much more common, and economical, to find a site and design a house to fit it rather than fitting the site to the house, so let's look at design from that point of view. It is of utmost importance to understand just what your site is all about.

Houses for Sloping Sites

Consider a house that is built into the side of a hill. Sloping lots offer the following design advantages: It is possible to have an exit at ground level; ventilation can be arranged fairly easily; drainage is usually good; frequently there's access to a view; the shape of the earth covering the house can be formed to fit the shape of the land; daylighting does not have to depend on skylights or an atrium, since at least one vertical wall is fully or partially above

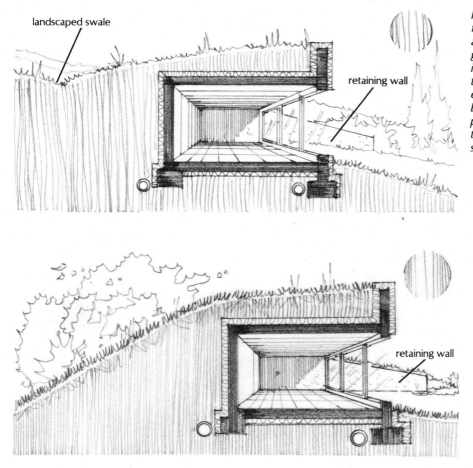

landscaped swale

retaining wall

Figure 2-1: The site is an important factor in determining the design of an underground house. Underground houses are typically built into a hillside or on a flat site, with the roof and three sides (north, east and west) covered with soil. Earth-contact houses have exposed roofs and are usually partially or totally bermed on three sides.

retaining wall

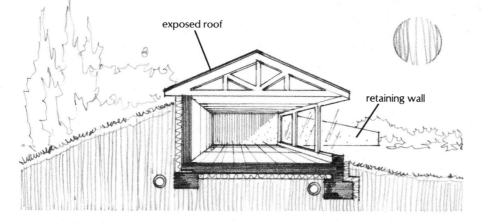

exposed roof

retaining wall

*Figure 2-2: For steep slopes, a
long, narrow house that is planned
to be parallel to the slope requires
the minimum of excavation, and
during excavation, there's less likelihood of striking water or rock.*

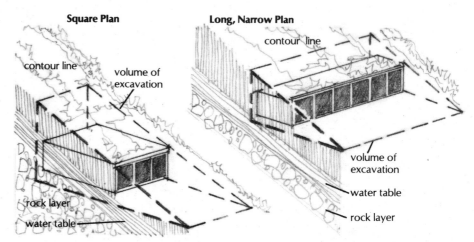

grade; at least one wall may be of lower-cost frame rather than masonry
construction; and, if the slope faces south, passive solar design is very easy to
incorporate.

The actual floor plan to use naturally depends a great deal on personal
preferences, but usually a relatively narrow house whose length parallels the
contour of the slope will provide the most house for the least money. This is
particularly true if the slope is steep — rising more than 4 feet for every 10-foot
horizontal run. With such steep slopes, two costly situations face the builder of a
house that extends far into the hillside. The first is that more earth will have to be
excavated as you go further into the slope. Also, you are more likely to run into
rock, which can become a very expensive proposition. Narrow houses require
the least excavation per unit of floor area, and that usually means the least cost,
both in excavation and in structure. Second, as the house is buried deeper, the
structural cost also goes up because of increased loads on the roof and
sometimes on the walls as well.

Excavating into a steep slope may also create a drainage problem on the
roof. If the roof cover slopes to the exposed (downhill) side of the house, the
parapet often used to keep the roof earth in place will also dam the water flow,
and while some designers include drainage paths through the parapet (called
scuppers), they are very prone to plugging with dirt or ice. Water will thus
accumulate on the roof, resulting in roof leaks. To keep that from happening, a
common design approach is to position the house in relation to the slope so that
drainage is to the back and the sides of the house. That means that the runoff
from the slope above is diverted before it gets to the house, and the only water
that reaches the roof is that which actually falls on it as rain. A steep slope could
be reformed to achieve this very useful effect.

There are basically four approaches to surface-water control for a house

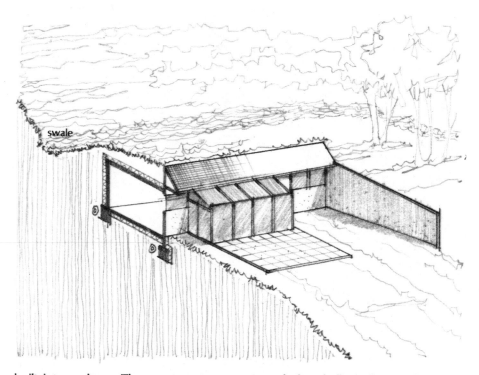

swale

*Figure 2-3: A swale behind an un-
derground house diverts water
from the roof and natural slope
around the house. A swale ensures that a minimal amount of
water will collect around a house,
but it is suitable only for shallow
slopes.*

built into a slope. The most common approach for shallow slopes, shown in
figure 2-3, is to have both the runoff from the slope behind the house and the
roof runoff drain to a swale behind the rear wall of the structure. The swale is
sloped to one or both sides and, if properly sized, keeps the slope runoff away
from the roof area. One problem with this design is that the swale should be well
away from the closest house wall to prevent the problem of water pooling at or
near the wall surface. I suggest that the base of the swale be no closer than 10
feet from the nearest house wall. Some designers place the swale close to the
wall but install a *French drain* to improve drainage. I have no argument with that
system, as long as percolation down to the French drain is rapid and as long as the
tiles at the bottom can carry away the water at an adequate rate. My personal
tendency, however, is not to depend too much on everything working as well in
20 years as it does when new. For that reason, I feel that the best approach is to
channel surface water away from the house before it soaks in. If the site will not
allow that, use the French drain approach. It's certainly better than nothing. If you
like having an extra insurance factor, do both.

In the second approach, illustrated in figure 2-4, the surface water from the
slope is diverted around the house so that the roof drainage system has to cope
only with water that actually falls on the roof. Like the swale method, the diverter

Photo 2-2: It is important that surface drainage paths be protected against erosion. This swale behind an underground house is maintained with a good sod cover.

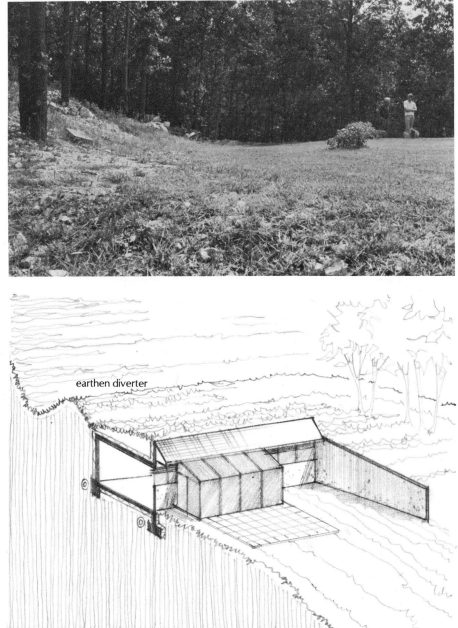

Figure 2-4: For steeper slopes or areas where large amounts of runoff may occur, it may be best to divert the water away from the house, being sure it is not dammed up.

earthen diverter

ridge should be at least 10 feet up-slope from any house wall. This approach is not as common, but it does allow a more even earth distribution on the house roof where slopes are steep or the roof area is large.

The third approach is to continue the slope to the roof line and, with a gutter or series of scuppers, to dump water over the front edge of the house (see figure 2-5). If the total watershed to the roof is not very large, that can work, but remember that scuppers do get clogged, especially in freezing weather. It also requires careful attention to grade on the roof, but the result does blend in well with the natural grades of the site. Be sure that the system used to get the water over the edge of the roof is large enough. Remember how full gutters on a conventional house can become in a cloudburst, even though they serve only the roof area. Your system may have to serve a runoff area several times as large

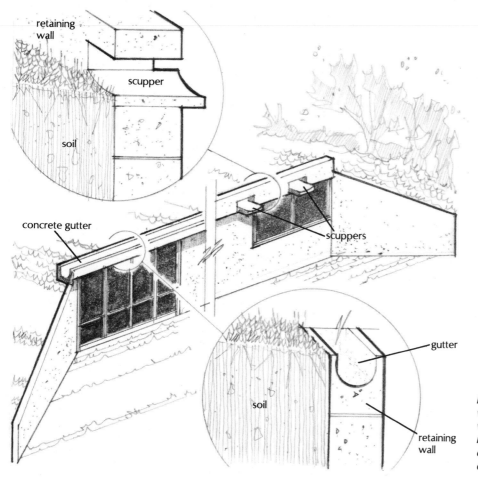

Figure 2-5: For those few sites where little runoff occurs and where ice and snow are not common, surface water may be run over the roof through scuppers or diverted to the side with a gutter.

Figure 2–6: A well-drained patio that intercepts runoff from a terraced, north-facing slope allows solar exposure while isolating a house from surface water.

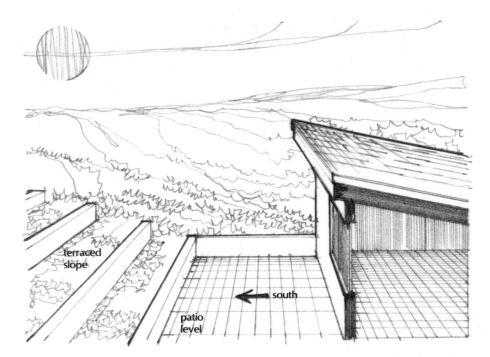

terraced slope

south

patio level

as the roof. In particular, I would be very hesitant about the "over-the-edge" system of roof drainage if the total watershed served is more than four or five times the actual roof area, *especially* if a large part of that watershed is rock, asphalt, concrete or bare earth, which encourages rapid runoff. Sod is much more forgiving.

Several houses have been built with the roof as a continuation of the hillside slope rather than the more conventional opposing slope (see figure 2–6). Properly done, this fourth approach can be excellent. Obviously, good drainage must be arranged at the patio level, but that is usually fairly easy. The roof lines are a continuation of the site's slope. Drainage leads away from the buried parts of the house, and if the slope is not severe, a good solar exposure can be had for a north slope without seriously disrupting the natural lay of the land. Note, by the way, the use of the shed roof here. A shed roof allows higher ceilings, higher exposed (preferably south) walls with better illumination and ventilation, superior drainage and even a lighter, less-costly structure because of the more even distribution of the earth on top. If a sod-covered wooden roof structure is used, a triangular truss is particularly suitable for this design (see figure 2–7).

Some lots will not readily accommodate a long, narrow house, or the builder may want a house with more than about 2,000 square feet of floor area, which would be awkward if the house were only 20 or 30 feet front to back. In such cases, the underground split-level approach can be used with success. The use of the split-level keeps excavation costs at a minimum, as well as allowing relative ease of daylighting in the back of the house, even though it is deep in the slope. If this design is used on a south slope, clerestory windows are a great advantage for both passive solar and daylighting. One must be careful, however, to allow good drainage from the lower roof. Drain tiles at strategic locations, such as the junction of the upper and lower parts of the house, make good sense. Generally, sloping the lower roof grade to the front of the house and directing the water through surface channels and drain tiles at the base of the parapet keep water from finding its way into joints or the fine cracks in the roof structure that will sooner or later occur.

Multi-story underground houses have been built, often into the side of a cliff, but suitable sites are rare, and such construction can be exceptionally costly if the site is not appropriate.

Where very steep slopes or actual cliffs are present, the two-story house can perform very well. Passive solar, in particular, is practical, since the front-to-back distance in a two-story house can be just one room deep. That permits maximum solar penetration to all rooms. Also, air circulation is easy to arrange in two-story designs, since the house is more compact and room interconnections are short and simple to arrange. An example of a successful two-story underground house is described in chapter 8. For the site selected, the two-story design was easily the best choice for this house.

Some sites are particularly suited to open-atrium houses. A U-shaped house can be outstandingly attractive and can perform well, provided there's good drainage. Beware, however, of the site that is similar to this one but where the U is in reality a water course with a watershed of several acres. You are supposed to be designing an underground house, not an underwater house! When the house is completed, the slope behind the structure should be short or sloping to the sides so that it doesn't funnel water right to the house roof. If there is a fairly long slope behind the house that would cause considerable water to flow toward the house, an earthen water diverter at least 10 feet from the closest wall can prevent this (see figure 2–9).

A site located at the top of a moderately sloping hill has excellent potential for earth sheltering. This location allows unusual freedom of orientation and is

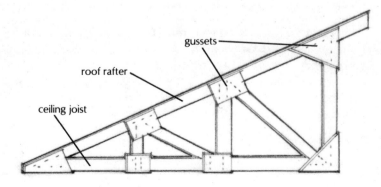

Figure 2-7: Economical triangular trusses can provide high strength for a wood-framed shed roof.

Figure 2–8: A split-level design, which can be built with minimal excavation, allows increased day-lighting and greater potential for passive solar heating.

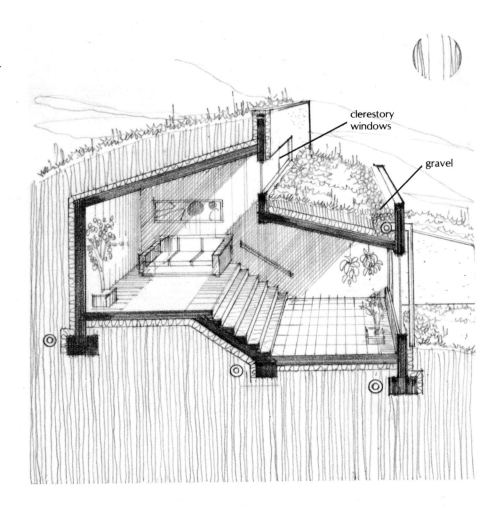

clerestory
windows

gravel

excellent for water control, since there is almost no chance of standing water being in contact with below-grade surfaces. This site can provide an excellent view and ease of ventilation, and it allows for both front *and* back doors. All in all, it is my favorite type of site.

You do *not* want to place an underground house at the base of a hill or in the bottom of a valley. It will be very difficult indeed to prevent water intrusion and erosion problems where you have modified the existing lay of the land. If you have a valley site and really want to build there, be sure you visit it right after a heavy thundershower so you can see the huge amounts of water that will have to move around (not over or through!) your house.

Building into Flat Sites

My definition of a flat site is one that has less than 4 feet of vertical drop over

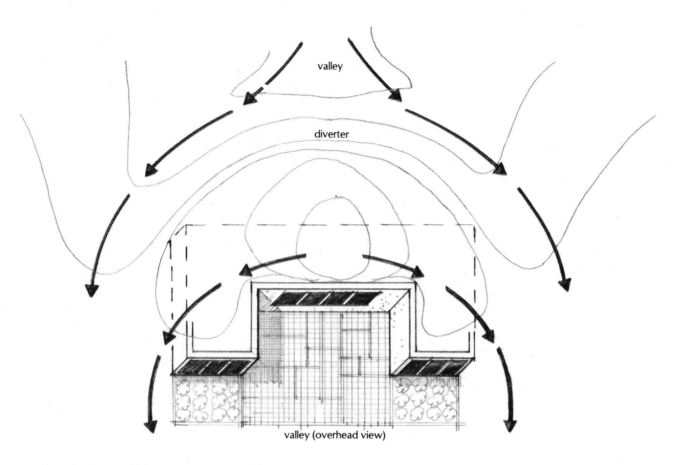

valley

diverter

valley (overhead view)

a radius of at least 100 feet. Many parts of the country have soil deep enough to allow you simply to dig straight down 10 feet and bury the house, but it is rarely a good idea. The main problem is water removal from the site, and the secondary problem is access both for people and for daylight. Another problem, which may sound trivial but certainly is not, is what to do with all the dirt you take out of the hole! Hauling it off can be very expensive, since a typical house will displace around 700 cubic yards of the stuff. That is close to 100 trips by a moderately sized dump truck! Still, the real problem is water. As I have noted before, and will again, by far the best way to waterproof a house is to keep the water away. That means drainage tiles all around the footings and under the floor. Drainage tiles are supposed to drain, so in order for gravity to do the work, there must be an outlet to the surface that is lower than the collecting tiles. In a house buried 10 feet deep into a flat site, that may be tough to find. As a negative example, one inexperienced underground builder remembered that one way to get rid of water from gutters was to dump it into a *dry well,* which is a pit containing gravel.

Figure 2-9: An open-atrium house fits well into a shallow valley, allows an excellent view and requires minimal excavation. Water must be diverted around the house, however, if there's a significant upslope watershed.

If it works for gutters, he thought, why not for drain tiles? So, at a depth of 10 feet, he built several dry wells and led the drain tiles to them, expecting this to drain water away from the house. What happened, of course, was that during the rainy season the groundwater level rose above the outlets of the drain tiles (the ''dry wells' were very wet by then), and the house was literally immersed in water several feet up the walls, and there were leaks! Would a sump pump help in this situation? It would as long as the house had power, but massive electrical storms that dump inches of rain often cause lengthy power outages. A better solution is to figure out a way to run the drain tiles to a surface outlet. What about a ditch meandering near the house? That's fine as long as heavy rains don't fill the ditch above the outlet level of your drain. If they do, water may flow back to the house instead of the other way around.

So what *can* you do with a flat lot? Simply put the house into the ground a few feet, and pile soil onto the roof and against three walls to create a pleasant little hill on a formerly flat site. A variation of this design is the *earth-contact* house, which has a fully exposed, conventional roof and earth berms usually on three walls, either partially or completely up to the roof (see chapter 5 for design requirements of earth-contact construction).

Ideally, the highest spot on the lot should be the house site so that the floor

Figure 2-10: Below-grade drainage is very important to successful waterproofing. Drains should slope continuously toward the outlet. Underfloor drains may be needed in wet climates or when the span exceeds 30 feet. If you're building near a creek or stream, maximum floodwater height should, of course, be below the outlet of the drain.

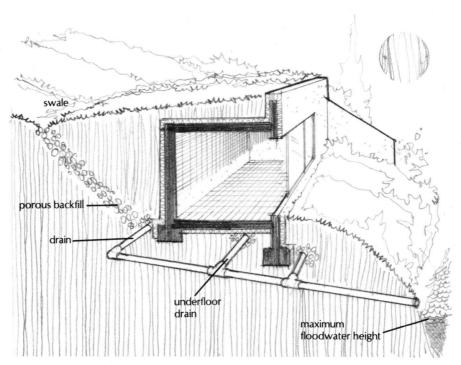

can be placed as deep as possible and still have successful footing drainage. With careful design, the amount of dirt removed from the house site can all be used to cover the house. That eliminates costly hauling charges and keeps the rich topsoil in your yard instead of in someone else's. Done properly, this site can have the same characteristics as a hilltop site: The house can have a solar orientation; drainage can be very good, since only the house area is the watershed for the site; access for people and automobiles is easy; and the house can be just about any shape without serious cost penalties (unless there is a significant energy penalty because of inadequate solar exposure).

There is, by the way, another almost unbelievable disadvantage of putting the underground house deep into a flat lot: It may float! A house that displaces 700 cubic yards of dirt will also displace some 1,180,000 pounds of water. That may actually be more than the weight of the structure before adding the roof and backfill, and if it is, the house will have the ability to float. A few years ago, I probably wouldn't have believed it, but that was before the case of the Floating Swimming Pool. A motel owner had an indoor swimming pool, and with winter business way below that of previous years, he decided to drain the pool for the winter to save the cost of heating the pool house and operating the pool system. Then came the spring rains. The groundwater rose to within a few feet of the surface, and the pool's entire shell, a quarter of a million pounds of concrete, rose majestically from its hole, slowly tearing away piping and the surrounding walkways. (No one actually saw it happen, but moving that much concrete implies a certain innate majesty.) The pool then tilted and came to rest about 2 feet out of its bed.

Photo 2–4: An earth-contact house, which has a conventional exposed roof, may be easier for a do-it-yourselfer to construct and has most of the features of a fully underground house. Seventy percent of the walls of this earth-sheltered house are bermed, which helps to moderate indoor temperatures throughout the year. The primary source of heat is direct-gain solar energy through south-facing front windows, two sloping windows and clerestory windows. A 4-inch-thick concrete slab floor in the kitchen and living room acts as thermal storage mass. The solar performance of the house is enhanced by tight construction that prohibits air infiltration, insulation levels of R-38 in the roof, R-25 in the walls and R-5 under the floor, and movable insulation for the windows. Two recessed windows on the north side of the house (not shown in the photograph) allow summer ventilation and permit emergency egress.

I, too, have had problems with floating concrete. One time, after working for two days digging a hole and then placing a concrete septic tank in it, I returned the next day to find that the hole had filled with water, which floated the septic tank and caused it to come to rest in the most disadvantageous position possible. So, yes, Virginia, concrete can float, and although I have yet to hear of an underground house leaving its moorings, I will no longer be surprised if one does. If it is yours, please allow me a precognitive "I told you so."

Atrium Designs

One feature that has been included in many underground houses is the atrium. Basically, it is an open area surrounded by house, like a small courtyard. One of the reasons for its popularity in underground design is that it allows floor plans with areas in excess of the usual 1,500 to 2,000 square feet to have access to the outdoors from all major living spaces. An atrium can provide ample

Photo 2-5: This ancient Roman underground house, Maison de la Chasse, in Bulla Regia, Tunisia, features an atrium.

opportunity for daylighting, ventilation and even a little extra solar energy. On the other hand, it also increases the exposed surface area of the house, which can increase infiltration, heat loss and summer cooling loads. For the best energy performance, the atrium should be rectangular, with the long axis running east and west and with a separation of the north and south walls of no less than about 15 feet and no more than about 25. If it is less than 15 feet, the glazing that can capture solar energy will be shaded too much of the day. If the long axis is north and south, the south-facing area is reduced and the less desirable east- and west-facing areas will be increased. The same basic rules of window placement should be observed for an atrium as for windows around the outside of any house: Minimal window area on the east, west and north; all windows with multiple (double or triple) glazing; and as large a window area as practical facing south. The design attitude should be that windows on all sides except those on the south for solar gain are simply there for a little daylighting and ventilation. Excessive window area facing east or west will add much too much heat in the summer, and all but south windows will lose more heat than they gain in the winter. Rarely will it be useful to have the window area exceed 10 percent of the adjacent room floor area, except for south-facing walls.

Because the atrium is surrounded by the house, separate drains must be provided for it. There should be one drain for each 100 square feet of atrium. In no case should roof drains be directed into the atrium, even if it has good drains of its own. If the atrium drain system should become clogged, a flood could result.

Photo 2-6: When properly de-signed for a site, an atrium in an underground house provides light, ventilation and outdoor living space, as in Ecology House.

You should be aware that with certain wind conditions, the atrium can become a very turbulent place. Try to slope the roof soil away from the atrium as smoothly as possible, especially on the side toward the prevailing winds. If there is a high parapet or other abrupt obstruction between the wind and the atrium, a turbulent swirl can develop that can carry snow and rain into the atrium in greater quantities than that falling out in open spaces. Also, sitting in such a space during a brisk spring breeze may not be as pleasant as you had hoped when it was designed.

Keep in mind, too, that if there are lots of deciduous trees around the house, the atrium will tend to fill with leaves in the fall, and bagging them up and carrying them through the house for ultimate disposal is not always a pleasant chore.

For some designers, the solution to these problems is to cover the atrium with a glass dome or series of plastic panels. In winter, that may be a good idea (as long as the structure meets building codes and is strong enough to handle the snow loads), but for the rest of the year, the heat collection will be too great for comfort, and unless large ventilating surfaces and/or external shading are included, the atrium may be totally unlivable in the spring, summer and fall whenever the sun is out. I strongly urge you not to consider an atrium as a fledgling greenhouse but rather as an open space. Build your greenhouse into the atrium if you wish, but follow the rules presented in chapter 3 to prevent the house from overheating.

A pseudo-atrium or open-end atrium will fit into many designs. Here again, try to minimize east and west surfaces and to maximize southern exposure. If the opening is to the north, it's still important to minimize the east and west glazings, since they are the prime culprits in increasing air-conditioning loads. Excessive north window area will also increase heating loads. The colder the climate, the more carefully these window areas should be considered in the design.

What about Rocks?

Striking rock can be both a help and a horror. In can be a help, since foundations placed on rock will be far less likely to settle than will those on soil. Rock can be a horror if it protrudes into the space the house must occupy. Blasting is very expensive, and the neighbors may not speak to you for quite a while afterwards. Rock usually follows surface contours, so if you run into rock early in the excavation of a sloping site, it will probably get worse instead of better as you dig into the hill. Many times, particularly with limestone or other sedimentary rock, a large bulldozer with an attachment called a ripper can cut the rock without blasting. If you have the misfortune to run into granite, basalt or other dense, hard rock, blasting will be your only recourse — if you insist on

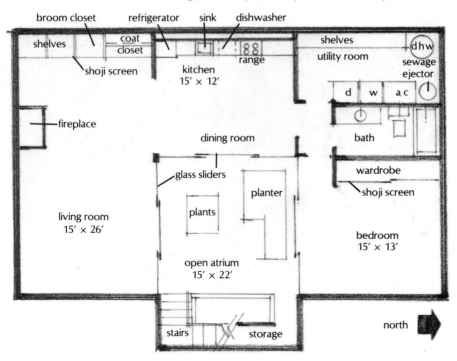

Figure 2-11: Floor plan of Ecology House.

SOURCE: Redrawn from plans with the permission of John E. Barnard, Jr., A.I.A., Marstons Mills, Mass.

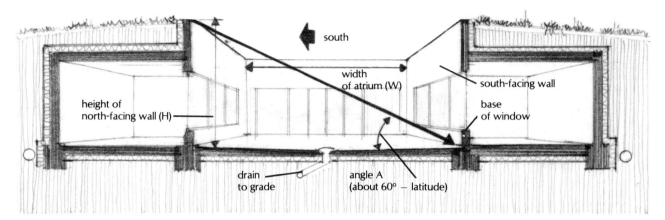

Figure 2–12: The angle between the bottom of south-facing atrium windows and the roof line of the north-facing wall should be less than approximately 60° minus the latitude if solar heating capability is desired. To determine this angle (A), divide the height of the north-facing wall (H) by the distance between the north- and south-facing walls (W). The following provides the approximate angle:

RATIO H/W (TANGENT OF ANGLE)	ANGLE A
1.0	45°
0.9	42°
0.8	39°
0.7	35°
0.6	31°
0.5	27°
0.4	22°
0.3	17°
0.25	14°

It's also important to consider the length (L) of the south-facing wall of the house and the width (W) of the atrium if solar heating is desired. For latitudes from 30° to 35°, $\frac{L}{W}$ should equal about 2½; from 35° to 40°, $\frac{L}{W}$ should equal about 3 or more; from 40° to 45°, $\frac{L}{W}$ should equal about 4 or more.

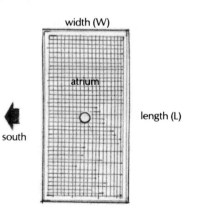

Photo 2-7: An open-ended atrium design provides most of the advantages of a closed atrium while allowing increased solar exposure and access to a view.

digging further. Obviously, in many cases, one easy way out is to dig to the rock and use that as the footing grade, build the house and then bring in more dirt to cover it. Unless the amount of dirt required is enormous, it will probably be cheaper than blasting and hauling away the rubble. One point of caution, though: It is generally not good practice to have one part of a house with footings on solid rock and another part on soil. Such a situation invites differential settling (more settling in one part of the house than another), with cracks and structural distortion being the results. In most cases, if the rock is at floor grade in one place, it will be within a few feet over the whole house site, so the proper course of action is to dig a trench for the footing down to solid rock so that the whole house is set on rock instead of just part of it. The floor slab can be on soil, since the load is carried by the footings.

As a part of the site survey described in chapter 4, determination of the location of rock layers is important. Designing to take advantage of rock instead of fighting it is then possible.

Laying Out the Floor Plan

An underground house requires a few special considerations in room layout. Since underground houses usually don't have attics or basements, all storage areas must be included in the living level(s) unless an additional building (usually a garage or storage shed) is built on the surface. Special attention must be given to the placement of closets and built-in cabinets as well as to possible locations for

Photo 2–8: Heavy-duty excavating equipment can often remove rock without the need for blasting.

storage rooms to house the domestic water heater, the heating system and other mechanical equipment. In the case of houses that do not face south, storage and utility rooms may be added as attached aboveground structures at lower cost than the cost of fully underground rooms. These "add-on" structures may actually improve the appearance of the exposed facade by giving it a more three-dimensional look. The presence of the storage space between the outside and the house proper will be useful as a buffer as well.

A second floor-plan problem is the placement of more than one entryway. Usually, it is very difficult to provide both a front and a back door to an underground house, although the atrium house and the hilltop house can be designed to do this. The main reason for having multiple entrances into a house is that they also serve as multiple exits in case of fire or other emergencies.

To be useful as emergency exits, entryways should be placed so that a fire in a particular location will not block all exits. It is not uncommon to place the kitchen at the junction of the bedroom hallway and the living room. If all the exits are at the end of the house opposite the bedrooms, a fire in the kitchen — one of the more common sites for such an emergency — can be a disaster. It is true that the house probably will not be consumed in flames, but it is also true that a fire in a closed place will quickly use up the oxygen in the house, and it will generate a lot of toxic smoke and gases. The actual damage to the house may be slight, but if

Photo 2-9 (left): *Since underground houses don't have attics or basements, storage space should be arranged for efficiency and maximum capacity.*

Photo 2-10 (above): *Alternate emergency exits from all rooms are desirable. Special crawl spaces, back hallways or in this case a passageway through adjacent closets are preferable to ceiling exits, since smoke and toxic gases accumulate at the highest parts of a room.*

the occupants can't escape in a few minutes, the results for them may be lethal. So, create your floor plan to provide convenience and allow for personal preferences, and then examine the plan with emergencies in mind and place roof exits, entryways, passageways and atria to ensure egress from all major sections of the house. Photo 2–10 shows an emergency exit through a closet.

The fact that there is no attic or basement in an underground house not only affects storage space allocation but also creates problems for the placement of ventilation and heating or cooling ducts. Although the ducts can be placed under the floor, it is usually an expensive process. For that reason, it generally saves money to place the house's mechanical equipment in a central location rather than at the far end of the house as is often done. Ceiling runs for ducts can often be disguised as beams, or, in halls, kitchen or bathrooms, the ceiling may be lowered to 7 feet instead of the usual 8 and ducting placed in those furred-down areas. If attention is given to room layout, it is usually possible to have all rooms adjoining a furred-down section, thereby allowing low-cost duct runs. One side effect of central placement is more noise in the main living area unless special attention is paid to insulating and soundproofing the mechanical room. Soundproofing room walls with insulation is generally a lot cheaper than buying long

Photo 2-11: When finished, this
ceiling will be furred down to con-
ceal heating and cooling ducts and
electrical and plumbing runs.

duct runs. The improved operating efficiencies of centrally located equipment
will also be a dollar saver.

Your plumbing system can also be arranged for efficiency and economy.
Because plumbing is expensive, placing all rooms that require water and drains
close together is a very good idea. Keeping hot water runs short not only saves
pipe, but it stops the long waits for hot water to appear after turning on the tap.
When there's more than one bathroom, placing the two water closets back-
to-back across a dividing wall can save hundreds of dollars over other alternative
locations. In addition, stoppages are less likely, and if they do occur, they are
more easily cleared.

As a final thought, as you draw your floor plan, do your best to picture it as it
will be built and finished. Think about where you will place your furniture. Think
about where people will walk. Consider the placement of doors and their
direction of swing. Don't do as I did in one house: I placed a door into a small
bathroom in such a way that the occupant practically had to climb onto the toilet
seat to close the door. If I had heeded the advice I so freely give and just thought
about it before I built it, I would not have had the embarrassment and cost of
redoing that doorway.

Getting Something for (Almost) Nothing

Throughout the design process of your underground house, think about
ways to make components do several jobs. Locate interior walls where structural
supports are needed. Concrete wall sections that are needed to hold up the roof
can also serve as solar heat storage walls. Ducts needed for heating can double as

ventilation passages for cooling. Anywhere that more than one function can be combined into a single component, money is saved, and your design is more efficient. Remember that what you design will literally be cast in concrete. Changing the design while it is on a piece of paper is nothing compared to trying to modify a completed underground house. Get it right on paper. If you are not supremely confident in your design ability, get a second, a third or even a fourth opinion. Don't hesitate to pay for at least one of those opinions, since all too often free advice is worth just what you pay for it. A hundred dollars or so spent during the design stage can save you thousands later.

Making a model of your design, showing especially the interior arrangement, can be vastly superior to a two-dimensional plan for visualizing the way a particular room arrangement will work. Architects and engineers who work with plans all the time have models built costing thousands of dollars to help them foresee three-dimensional problems that are almost impossible to visualize from two-dimensional drawings.

Summary of Underground Design

The real key to successful overall underground house design is working with the site in mind. No matter what, arrange for drainage away from the structure both at the ground level and down at the footing level. Generally, if the structure is laid out with the long axis parallel with the site's contour lines, you will get the most house with the least excavation cost. Don't be afraid to build the house mostly aboveground and then cover it up with berms. There is no underground building oath you have to take that requires you to have all of the house below the natural grade.

When designing the details of the plan, try to balance the requirements of traffic, furniture placement, ventilation, emergency egress, lighting, plumbing, heating, cooling and storage while retaining your sense of humor and your general sanity. Perhaps most important: *Get an outside opinion,* preferably an opinion from someone experienced in *successful* underground design.

CHAPTER 3

Solar Solutions for Underground Houses

"It is criminal to ignore passive solar techniques in the design of underground homes," says Lon Simmons, a prominent Midwestern builder of earth-sheltered houses. "The combination of passive solar and underground construction is one of those rare occasions where the sum is greater than the parts." While some people might argue that the word criminal is a bit strong, most people who understand solar design agree that most underground houses have much greater potential for passive solar utilization than just about any other house style. That potential does not become available automatically, however. Just as plumbers or electricians must have an understanding of the physical laws affecting the components used in their trades, so must the building designer be aware of where the energy comes from, where it goes and how a building interacts with the sun.

The Solar Source

The sun radiates energy over a wide spectrum, but predominately in the light wavelengths spanning the ultraviolet through the infrared. Figure 3–1 is a graph of the energy intensity of solar radiation as it leaves the sun. Notice that the curve has its peak in the visible part of the spectrum, but that as far as total energy is concerned, the infrared contributes about as much as the visible. The ultraviolet portion of the spectrum, although important to life, is relatively unimportant as an energy source.

Of course, the amount of energy leaving the sun's surface each second is staggering, but by the time it travels the eight minutes to Earth, energy levels available for our use have fallen to about 429 Btu per square foot per hour (0.135 watts per square meter). That is the level of energy received by the solar cells of earth satellites. Also, unless the satellite passes through the earth's shadow, that energy will be received continuously. At the surface, of course, we're not so lucky. We can't avoid being in the earth's shadow sometime during its 24-hour rotation. The sun goes down for the night (which makes it the coldest time of the day just because the sun *is* down), and in winter, it stays above the horizon for less time than it does in summer. Shorter winter days are the result of the earth's revolution around the sun each year. The earth is tilted on its spin axis. That causes the northern hemisphere to be tilted toward the sun part of the year (in summer) and away for another part (in winter). During spring and fall, little tilt is apparent.

The tilt of the earth's axis has an effect that is very important to solar design: The apparent path of the sun changes markedly as the seasons change. Thus, there are two important relationships to be

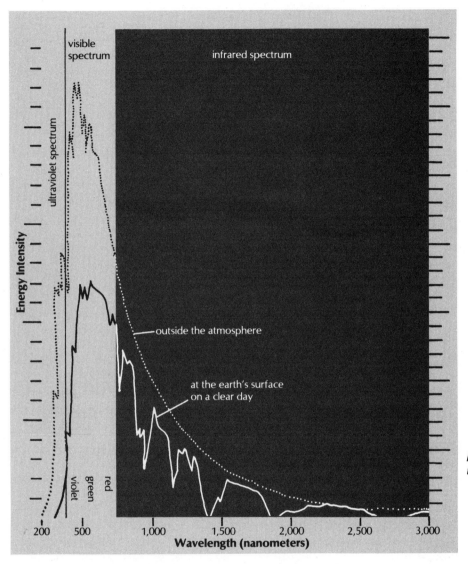

Figure 3-1: Graph of spectral distribution of sunlight.

Source: Adapted from Edward
Mazria, *The Passive Solar Energy
Book,* copyright © 1979, with
the permission of Rodale Press,
Emmaus, Pa.

aware of at your building site. First, the sun appears higher in the sky (solar altitude) for any given time of the day in the summer than in the winter. Second, the points on the horizon where the sun rises and sets (solar azimuth) vary considerably with the seasons. In the winter the sun rises and sets in the southern half of the sky, while in the summer it rises and sets in the northern half. If you want to ask your friends a trick question that is more original than "Do you still beat your wife, husband or dog?" ask them "How often each year does the sun rise in the east and set in the west?" The answer, of course, is twice. Once at the

*Photo 3-1: Because of the modest space-heating require-
ments of an underground house, solar design features can be
attractive and unobtrusive and can provide a large percentage of the heating needs.*

spring equinox (around March 21) and once at the fall equinox (around
September 23), when day and night are equal in length. The rest of the days of
the year the sun rises to the north or south of east and sets to the north or south
of west. On the longest day of the year, June 22, the rise and set points are at
their maximum distance north of an east-west line. By December 22, the shortest
day of the year, they are at their maximum distance south of that line. The earth's
axial tilt of 23.5 degrees causes this phenomenon. Fortunately for everyone,
these yearly and daily cycles are predictable. Solar designs can be made that
allow for, or even rely on, these annual changes in the path of the sun. Other
modifiers of solar energy are not so predictable.

    As sunlight (we can call all of its components "light" — that includes
ultraviolet, visible and infrared wavelengths) passes through the atmosphere on
its way to the surface of the earth, it is changed in several ways. These changes
come about through the interactions of the solar energy and the various solid,
liquid and gaseous components of the ocean of air surrounding the earth. The
first significant change that occurs — much to the benefit of life on earth — is the

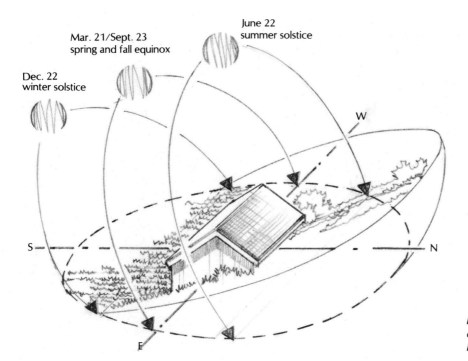

Dec. 22
winter solstice

Mar. 21/Sept. 23
spring and fall equinox

June 22
summer solstice

W

S

N

E

Figure 3-2: Positions of the sun during the year in the Northern Hemisphere.

absorption of the majority of the ultraviolet light by atmospheric gases. This absorption results in the production of a layer of ozone that is from 9 to 18 miles up. From there down to about 7 miles above the earth's surface, there is little additional modification of the sunlight. It is in the last 7 miles (called the troposphere) that most of the water vapor, clouds, pollution, dust and, in fact, air reside.

The gases of the air itself (78 percent nitrogen, 21 percent oxygen and 1 percent combined amounts of water vapor, carbon dioxide, argon, neon and trace amounts of many other gases) affect the sun's light in two ways. First, they selectively absorb some wavelengths and pass others unchanged. Second, they scatter the short waves (violets and blues), while having little effect on the longer visible and infrared wavelengths. Figure 3-1 shows the result. The deep valleys in the curve are the result of the selective wavelength absorption of atmospheric gases. The infrared portion of the spectrum is modified dramatically due to selective absorption by carbon dioxide and water vapor. The overall reduction in intensity is largely the result of molecular scattering by the air. Since scattering is nondirectional, part of the scattered energy is directed toward the earth (causing what we see as a blue sky), while the rest is scattered away from the ground and is lost to space.

The liquids and solids in the atmosphere have an even greater effect on the light that finally makes it to the lower atmosphere. The most common modifier is

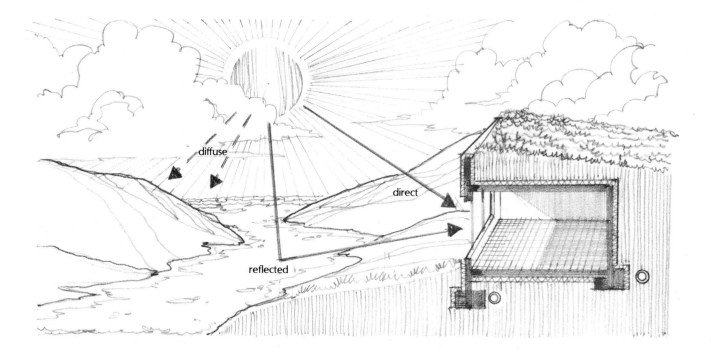

Figure 3-3: Solar radiation that reaches the earth is either direct, reflected or diffuse.

water droplets, that is, clouds. Solar energy levels may be reduced to less than 10 percent of the clear-sky values by thick clouds. Much of that reduction is due to the reflection of sunlight from cloud tops. If you have flown in a jet from a cloud-covered airport, you were almost certainly impressed by the brilliance of the clouds after you broke through to clear daytime skies.

Dust from wind erosion, human industry and volcanoes is another solar energy reducer. In cases where the dust is fine enough, it scatters light just as the air itself causes scattering. Large dust particles actually absorb solar energy, but most of the time, for most sites, dust is not a significant modifier of solar energy when compared to the effects of clouds.

The unpredictable nature of clouds makes it difficult to design the most cost-effective solar system possible. About the best a designer can do is to use historical weather data to average the cloud cover over a number of years to predict net solar energy "income" for future years. Fortunately, the cost of not providing the absolute optimum solar design is small when the overall energy demands on the system are small, so for our very energy-efficient underground house, it is not quite so critical to be able to predict energy needs and energy availability as it might be for a house of conventional construction. Figure 3-4 shows climate zones in relation to suitability to earth-sheltered construction. Although the information in this illustration is general and should be considered only as rough approximations of climate data, it is useful in combination with other aids in designing a passive solar underground house. Specific climate data and the latitude of a site have major effects on several passive solar design parameters. What that means, of course, is that universal designs do not make good economic sense. A builder who uses published, nationally distributed house plans without modification for local conditions should not be the one you hire to construct your solar dream house. A house that performs great in the mountains of Nevada may not work well in upstate New York. One that works fine in upstate New York will probably be a lot less than optimum for the Nevada site. The concept of site specificity is so important, let's interrupt at this point to state a rule of passive solar design: *Design to the site.*

Passive Solar Design

Since definitions for passive and active designs are not yet firm in the mind of everyone, the first order of business is to define what is meant here by passive solar energy utilization. It is the capture, storage and distribution of solar energy using integral building components and without using significant amounts of utility-supplied power. The main advantages of passive systems over active ones are cost and reliability. The lower cost is due to the use of the solar components for other than solar jobs. While passive solar components double as walls, floors, ceilings, windows and fireplace facings, active components are usually purely solar and therefore cost more for each Btu of energy delivered. Passive systems have few, if any, moving parts and are composed of long-life building materials, so there is little that can fail. Power outages, brownouts, low gas pressures and so forth do not affect systems that have no utility connections. For both these reasons, passive systems are outstandingly reliable. If there is a significant disadvantage to passive systems, it must be that they are harder to design for high percentages of solar utilization than are active systems. This is true because the occupants live *within* the solar system, so conditions must stay within a fairly narrow range of temperature and humidity, or discomfort results. Also, once you have built a passive system, it is difficult and expensive to modify because it is an integral part of the building. Active systems can have design errors corrected by simply changing components.

The basic design problem for passive solar space heating is not complex. The solar heater must provide a way for the solar energy to enter the house and then keep it there while maintaining the occupied space within the comfort range of temperature and humidity.

Active systems, those with discrete solar components in the form of flat plate collectors, storage units and solar heat transfer systems, are usually not even considered for underground housing. First of all, the most expensive part of a passive system—the storage mass—is usually abundant in underground houses. Second, the economics of active systems are particularly poor for small residential systems, and the great energy efficiency of underground houses

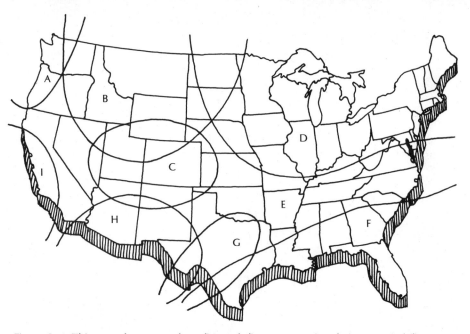

Figure 3–4: This map shows rough outlines of climate zones in relation to suitability to earth-sheltered construction.

In Zone A, cold, cloudy winters maximize the value of earth-sheltered construction as a heat-conservation measure. Cool soil and dry summers favor subgrade placement and earth cover, with little likelihood of condensation problems.

In Zone B, severely cold winters demand major heat-conservation measures, even though more sunshine is available here than on the coast. Cool soil and dry summers favor earth-covered roofs and earth-contact construction.

In Zone C, good winter insulation offsets the need for extraordinary winter heat conservation, but summer benefit is more important here than in Zone B. Earth cover is advantageous, and the ground offers some cooling; condensation is unlikely, and ventilation is not a major necessity.

usually means that a small system is all that is needed. Finally, underground houses usually have no conveniently sloped roof for collector installation, nor do they usually have a basement or attic in which to place other necessary components. Expensive, prime living space must be used for the storage unit and its associated distribution system. So, all in all, active space-heating systems are rarely the best choice, either technically or economically, for underground housing.

Siting and Orientation

Getting the sun into the space would be much easier if the sun stayed in the same place in the sky all the time. Since it does not, the sun's movement across the sky during its daily and annual cycles, and the way that movement affects the energy received on the various surfaces of a house, have to be carefully considered.

Look again at figure 3–2. In the winter, the sun is low and is always in the

In Zone D, cold and often cloudy winters place a premium on heat conservation. Low summer ground temperatures offer a cooling source, but with the possibility of condensation. High summer humidity makes ventilation the leading conventional summer climate control strategy. An aboveground super-insulated house designed to maximize ventilation is an important competing design approach.

In Zone E, generally good winter sun and minor heating demand reduce the need for extreme heat conservation measures. The ground offers protection from overheated air, but it doesn't have a major cooling potential as a heat sink. The primacy of ventilation and the possibility of condensation compromise summer benefits. Quality of design will determine the actual benefit realized here.

In Zone F, there are high ground temperatures. Persistent high humidity levels largely negate the value of roof mass and establish ventilation as the only important summer cooling strategy. Any design that compromises ventilation effectiveness without contributing to cooling may be considered counterproductive.

Zone G is a transition area between Zones F and H and has some of the qualities of both zones. Moving westward through this zone, the value of earth-sheltered construction increases, and moving southward, it diminishes.

In Zone H, summer ground temperatures are high but relatively much cooler than air. Aridity favors roof mass, reduces the need for ventilation and eliminates concern about condensation. Potential for integrating earth-sheltered construction with other passive design alternatives is high.

In Zone I, extraordinary means of climate control are not required due to relative moderateness of this zone. Earth sheltering is compatible with other strategies, with no strong argument for or against it.

SOURCE: Map redrawn and caption reprinted from Kenneth Labs, *Regional Analysis of Ground and Aboveground Climate* (Oak Ridge, Tenn: Oak Ridge National Laboratory, 1981). The original caption was slightly reworded.

---

southern half of the sky, which means, of course, that collectors must have a southern exposure. The northern exposure is easily eliminated from consideration, since it never gets any solar energy in the winter. East and west surfaces will each be illuminated half of the day, but during the early morning and late afternoon when the sun's rays can best penetrate east and west windows, the available energy has been greatly reduced by its passage through long reaches of the relatively dirty lower atmosphere. Also, a real problem with east and west windows occurs in the summer because they are still receiving solar energy for half a day, which can create increased cooling needs. North-facing vertical surfaces receive summer sun too, but as you can see in table 3–1, the east and the west sides of the house receive by far the most. Thus only the south surface is useful in solar collection for space heating.

What about houses that cannot be faced directly south? As it turns out, deviations of up to about 25 degrees east or west of south won't significantly reduce winter solar performance. Deviations of over 15 degrees, however, may

Table 3-1

Clear-Day Solar Heat Gain for Three Latitudes (in Btu/ft^2)

(The heat gain is through vertical double glazing at various orientations and through horizontal double glazing. The heat gains listed in the following account only for the reflection losses from the surface of the glass. To account for absorption losses, reduce the values listed by 6%.)

32° NORTH LATITUDE

	N	NE, NW	E, W	SE, SW	S	HORIZONTAL
JAN.	152	166	574	1,146	1,560	1,288
FEB.	192	278	772	1,200	1,424	1,688
MAR.	240	433	904	1,116	1,034	2,084
APR.	302	636	997	955	600	2,390
MAY	396	789	1,040	823	422	2,582
JUNE	450	841	1,038	758	390	2,634
JULY	408	789	1,024	803	420	2,558
AUG.	320	636	968	920	582	2,352
SEPT.	250	426	864	1,067	1,000	2,014
OCT.	200	280	746	1,151	1,364	1,654
NOV.	154	168	567	1,125	1,528	1,280
DEC.	136	144	518	1,128	1,574	1,136

40° NORTH LATITUDE

	N	NE, NW	E, W	SE, SW	S	HORIZONTAL
JAN.	120	128	474	1,079	1,506	948
FEB.	164	215	666	1,180	1,502	1,374
MAR.	220	376	858	1,183	1,244	1,852
APR.	294	593	1,002	1,075	838	2,274
MAY	384	747	1,063	952	598	2,552
JUNE	446	816	1,083	894	528	2,648
JULY	398	749	1,048	931	586	2,534
AUG.	312	595	975	1,034	806	2,244
SEPT.	230	370	816	1,126	1,190	1,796
OCT.	170	218	642	1,129	1,436	1,348
NOV.	122	130	466	1,056	1,472	942
DEC.	102	105	393	1,007	1,434	782

48° NORTH LATITUDE

	N	NE, NW	E, W	SE, SW	S	HORIZONTAL
JAN.	82	84	320	894	1,284	598
FEB.	130	153	540	1,106	1,486	1,040
MAR.	194	318	795	1,218	1,386	1,578
APR.	280	551	994	1,177	1,060	2,106
MAY	394	736	1,105	1,091	828	2,482
JUNE	468	820	1,144	1,042	740	2,626
JULY	408	741	1,092	1,068	806	2,474
AUG.	300	553	965	1,131	1,018	2,088
SEPT.	206	315	753	1,151	1,318	1,522
OCT.	138	159	523	1,056	1,414	1,022
NOV.	86	87	317	875	1,252	596
DEC.	66	66	237	777	1,130	446

SOURCE: From Edward Mazria, *The Passive Solar Energy Book,* copyright © 1979, reprinted with the permission of Rodale Press, Emmaus, Pa. The information in this table was taken from computer studies by M. Steven Baker, University of Oregon, Eugene, Oregon.

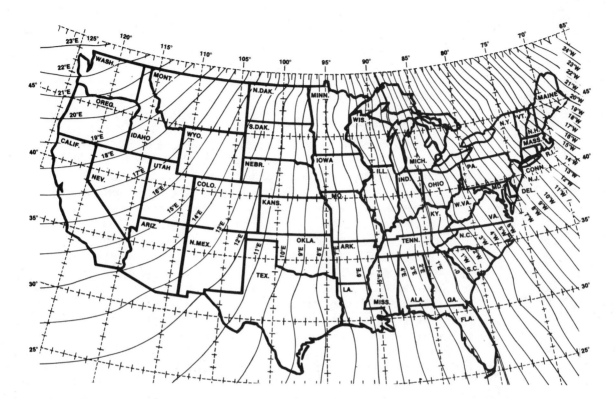

increase summer cooling needs. Beyond the 25-degree deviation, the amount of winter energy received is sharply reduced. A southeast face, for example, will have significantly increased energy collection in summer and considerably decreased collection in winter. But when you consider that a 25-degree deviation is acceptable, this means that orientation can vary by as much as 50 degrees (25 degrees east plus 25 degrees west). That's a lot of flexibility for any site.

For some locales, the deviation of magnetic compass direction from the true north/south direction must be considered. The map in figure 3–5 shows how that deviation varies across the United States and southern Canada. If you live in Vancouver, British Columbia, the true south will be 23 degrees east of compass south (180 degrees), which means that a true south reading on the compass is 157 degrees (180 degrees minus 23 degrees).

Figure 3–5: Compass deviation map.

Source: Redrawn from the Isogonic Chart of the United States, U.S. Department of Commerce, Coast and Geodetic Survey, 1965.

Direct-Gain, Indirect-Gain and Isolated-Gain Systems

There are three common approaches to the design of passive solar systems: direct-gain, indirect-gain and isolated-gain systems. These technical divisions are primarily for discussion purposes, since each category blends imperceptibly into the next. Isolated-gain systems blend into active systems, so even the difference between active and passive systems is blurred. In general, a blend of two or even all three of these systems can provide the best combination of cost-effectiveness and solar features. For the moment, though, look at each as if it were a totally independent entity.

Direct-Gain Systems

Most passive solar systems depend on a vertical or slightly tilted glazed surface as the means of getting the sun's energy into the house. The variables are due to what happens after the energy passes through the glazing. In the simplest passive solar method, the direct-gain system, the glazing surface is nothing more exotic than a window. Just any south-facing window does not a passive solar system make, however. The system must have the ability to get the sun's energy into the space and then keep it there while maintaining comfortable conditions. The key to successful direct-gain passive solar heating is the presence of systems that allow useful amounts of solar energy to come in the windows without overheating the space. For example, a bedroom space in a well-designed underground house may need only 20,000 Btu to keep it comfortable for 24 hours in 0°F weather. The difficulty is that 20,000 Btu received from the sun arrive over a 6- or 7-hour period, while the space needs heat for 24 hours — with maximum needs coming late at night. For those 6 or 7 hours, the room may be seriously overheated by a direct-gain system that provides 20,000 Btu during the day. After the sun sets, the room may get too cold. One solution is to provide a place where excess energy can be extracted from the room when potential overheating conditions exist, and then stored for release after the sun goes down. One such place is in large masses of masonry and concrete.

A large mass of concrete is certainly one thing most underground houses have. One pound of concrete will store about 1 Btu for every 5°F of temperature rise. So, if the room to be solar heated has a 10-inch thick concrete ceiling and a 4-inch concrete floor, for every square foot of floor area there are about 1.17 cubic feet of concrete available for storage on those two surfaces alone. That means that a 12-by-12-foot (144-square-foot) room would provide about 168 cubic feet of concrete for heat storage in the ceiling and floor alone, or about 24,192 pounds of concrete, since concrete weighs 144 pounds per cubic foot (144 × 168). About 20,000 Btu would raise the temperature of that mass only about 4°F. There are complications, naturally. Even though the mass is present in the room, there is no guarantee that most of the heat will end up in the concrete. The determining factor is the mass's ability to accept the energy at a rate comparable to the rate at which the energy is arriving. Anytime the sun provides energy at a rate faster than the storage mass can absorb it, the air in the room will overheat. To avoid overheating, the mass must be closely coupled to the incoming solar energy. Close coupling means that there is a minimum of conversion stages between the incoming energy and the resulting increased temperature of the storage mass. By far the best way to obtain the close coupling is to be sure that the sunlight falls directly on the surface of a dark-colored storage mass. This mass can be called *primary storage* when it receives at least four hours of direct sun each clear day. The sunlight-to-heat conversion takes place on the surface of the mass itself, and the heat transfer to storage occurs by conduction through the mass.

Unfortunately, about the only surface of a room that is consistently in direct sunlight from a south window is the floor. Exposing the floor to the sun means that it can't be shaded by furniture; nor can it be covered with carpet or other insulating materials. But it usually isn't possible to leave an entire floor clear of rugs and furniture. Enter *secondary storage*. This is mass that is available for storage but is not in the path of the sunlight. It can be used to store heat, but not as effectively as primary storage. For a secondary storage mass to accumulate heat from incoming sunlight, energy must be reflected onto it, or the air in the room must first be heated and then the heat in the air transferred to the mass. In most rooms, reflected energy provides little heat to secondary storage mass. For any given heat transfer system, two parameters determine the amount of heat transfer and the rate at which it takes place: the difference in temperature between the "sender" and the "receiver" of heat, and the amount of surface area available for transfer. In a direct-gain system, the maximum temperature of the air is limited by comfort requirements to about 80°F, and the lower limit of temperature for effective storage is likewise limited by comfort requirements to about 65°F. So the maximum temperature difference is 15°F. That turns out to be a problem because air is not only a poor heat transfer fluid, but it tends to form a stagnant boundary layer at the surface of the storage mass that further slows heat transfer. Since we are stuck with a small temperature differential and with a poorly performing heat transfer system, large surface areas are needed. But how large is "large"? At least ten times as much mass surface area is needed to transfer comparable amounts of heat to secondary storage mass as to primary storage. Some passive designers recommend that the secondary storage surface be as high as 40 square feet per square foot of south glazing to be equivalent to 1 square foot of primary storage. The moral: *Maximize primary storage.*

Some Direct-Gain Guidelines

With those relationships in mind, we can make reasonable estimates of the appropriate sizing of direct-gain systems for a very energy-efficient underground house. To do that, we need to know the size of the room being heated and its heat load under winter design conditions. The heat load can be calculated as per the method in Appendix 1. The size of the room is taken from the floor plan of the house. When the word "room" is used, I mean the total space that is bounded by walls, floors and ceilings. That may sound terribly obvious, but in the case where a kitchen, living room and dining room are all combined with no floor-to-ceiling walls dividing them, the room is the total floor area of all three

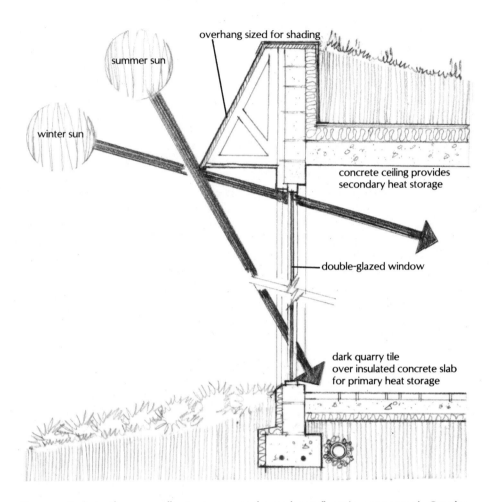

overhang sized for shading

summer sun

winter sun

concrete ceiling provides
secondary heat storage

double-glazed window

dark quarry tile
over insulated concrete slab
for primary heat storage

*Figure 3-6: A typical direct-gain
system.*

"rooms," since they are all one space as far as heat flow is concerned. On the
other hand, a bedroom adjacent to a large bathroom and connected by a door
that is usually closed would be considered two rooms for our purposes.

For this "quick and dirty" design method, the first thing to determine is the
maximum south window size that can be accommodated by the available
storage mass, both primary and secondary. To do that, first allow 15 percent of
the room floor area as "free." This means that a room without any specific
storage mass could have a south window area of up to 15 percent of the room's
floor area. Next, for each square foot of primary storage surface area, you can
add 1 square foot of window. Finally, for each 10 square feet of secondary
storage surface area, add 1 square foot of window.

In the sample room shown in figure 3-7, 15 percent of the 100-square-foot
room (10' × 10') is 15 square feet. The primary storage is the 2-by-10-foot band

of directly exposed floor behind the south wall, which allows an additional 20 square feet of window (assuming the window allows the sun to shine on all of that band). The secondary storage is the ceiling and the west wall (both of plastered concrete), for a total secondary storage surface of (10′ × 10′) + (10′ × 8′) = 180 ft². That allows another 18 square feet of window. Thus the largest window meeting our rough-and-ready guideline is 15 + 20 + 18 = 53 ft². Based on this approach, we can expect that installing about 53 square feet of south-facing windows (plus or minus 10 to 15 percent) will result in little or no overheating in most climates, while providing cost-effective solar heating. In this example, please remember that the number 53 is not magic; it is just a guide that is reasonable for climates with 5,000 or more annual heating degree-days (HDD). (A heating degree-day is a unit that represents the difference in °F between the average daily outdoor temperature of a particular day and 65°F. The assumption is that space heating is not required when outdoor temperatures are 65°F or higher. For example, a day that has an average daily outdoor temperature of 35°F has 30 heating degree-days. The total number of degree-days for a heating season is used to calculate heating requirements.) In warmer climates, direct-gain system size may be limited more by the heating needs of the house than by the amount of primary and secondary storage available.

After figuring the maximum window size as determined by available

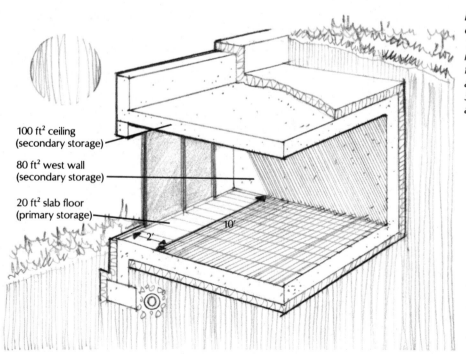

100 ft² ceiling
(secondary storage)

80 ft² west wall
(secondary storage)

20 ft² slab floor
(primary storage)

Figure 3-7: In the example of the direct-gain window sizing method, 100 ft² of floor area, 20 ft² of primary mass storage area and 180 ft² of secondary mass storage area allow for the installation of about 53 ft² of south-facing window area.

storage, you can plot a graph (see figure 3–8) to arrive at a maximum direct-gain window size as determined by the room's actual heat load. Using the estimated design-day heat load for the room (from the method in Appendix 1), draw a horizontal line from the column at left, which is marked "Design-Day Heat Load and Net Solar Gain (Btu/day)." Using the model room described above, the design-day heat load is calculated as follows (from the method in Appendix 1):

The parameters include a design temperature of 10°F, a Home Heating Index (HHI) of 5 ($Btu/ft^2/HDD$) and a heated floor area of 100 ft^2:

$$(65 - 10) \times 5 \times 100 = 27,500 \text{ Btu/day}$$

Thus, for the example, the horizontal line is drawn from the point corresponding to 27,500 on the vertical axis of the graph.

The next line to draw represents the maximum direct-gain window size that was calculated on the basis of storage availability. Draw a vertical line from the horizontal axis of the graph, marked "Window Size (ft^2)," starting from a point that corresponds to the number you calculated. (In the example graph, we're still using the 53-square-foot result.) The reason that the load goes up as window size increases is that the window allows much more heat to escape than does the solid, insulated wall. So, for every square foot of poorly insulated window that's added, a square foot of well-insulated wall is subtracted.

Now we look at the amount of solar gain that will be admitted by a given area of south glass. This graph was developed using a reasonable estimate of the solar radiation passing through vertical glazing on a sunny winter's day (1,000 Btu per square foot per day). It's true, of course, that this gain varies with latitude, and if you want to modify the graph to be more representative of your area, you can use the values for different latitudes that appear in table 3–1. Getting back to the example graph, notice where the design temperature line (slanted lines) and the room heat load line (horizontal lines) intersect. This intersection represents the maximum direct-gain window size *as limited by room heat load.* If we are working with a design temperature of 10°F, then we see from the example graph that the intersection occurs over the 88-square-foot point on the horizontal axis (window size). If, in actual design and construction, that window size were exceeded, there would be more solar heat gain than necessary for that room, even on the winter design-day (the coldest day). That, of course, means overheating, particularly on milder days or when several sunny days occur together. You can go beyond this recommended size if you want, but it will be necessary to provide a way to deliver heated air to other parts of the house, or to greatly increase the storage mass. What if the design temperature line never intersects the room heat load line? This means that the window size is limited by the available storage mass (primary and secondary) and not by the heat load of the room. Another question is, why the difference between the area allowed by mass and the much larger area allowed by heat load? In the simplest terms, it's

Figure 3–8: This graph represents average net solar gain for double-glazed windows at different outdoor daily average temperatures (based on window insulation value of R-2 and solar gain of 1,000 Btu/ft²/day). To use this graph, draw a vertical line representing the maximum useful window size that can be accommodated by available storage. (Refer to the sample estimation preceding this illustration.) All window sizes to the right of this line are too large for the available storage mass in the room and may result in overheating in fall and spring. Next, a horizontal line is drawn representing heat loss from the room (note that this line will move for changes in outside temperatures and in window size). Then, moving right to left, if the slanted solar gain line intersects the horizontal line before it intersects the vertical line (as the 10° line does in the example), then this intersection point, which, in the example, corresponds with about 88 ft², represents the maximum window size needed. (Note that the solar gain line you work with must correspond with the design-day temperature.) This means that the heat requirements of the room limit the need for large direct-gain solar window size. If the slanted solar gain line (the 10° line in the example) were to intercept the vertical line first, then the maximum usable window size would, in fact, be limited by the availability of storage. Additional (auxiliary) heat would have to be added in the amount of the vertical distance between that intersection point and the horizontal room heat load line. In the example that amount is labeled "extra heat required (Btu/day)," which in this case equals about 12,500 Btu. Again, window sizing by available mass is a way to prevent overheating during marginal heating periods (in spring and fall).

When you develop your own guidelines, you can use the parameters defined by the existing graph (solar gain of 1,000 Btu per ft² of window per day) or you can create your own parameters using the more precise solar gain information in Table 3–1 and the following formula:

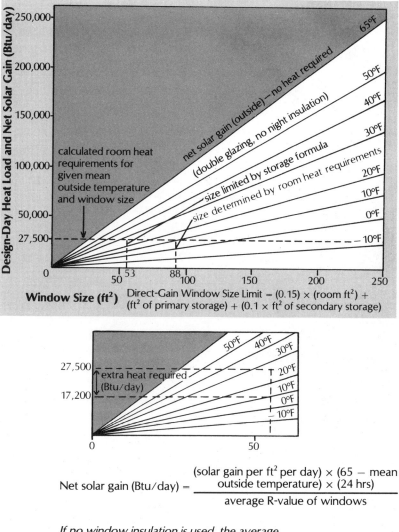

Direct-Gain Window Size Limit = (0.15) × (room ft²) + (ft² of primary storage) + (0.1 × ft² of secondary storage)

$$\text{Net solar gain (Btu/day)} = \frac{(\text{solar gain per ft}^2 \text{ per day}) \times (65 - \text{mean outside temperature}) \times (24 \text{ hrs})}{\text{average R-value of windows}}$$

If no window insulation is used, the average R-value to use is 2. If insulation is used for part of the 24-hour day, you can use the following formula to calculate an average R:

$$R_{average} = R_{glazing} + (R_{insulation} \times \text{fraction of the day insulation is used})$$

Say, for example, that R-8 insulation is used for 14 out of 24 hours:

$$R_{average} = 2 + \left(8 \times \frac{14}{24}\right) = 6.67$$

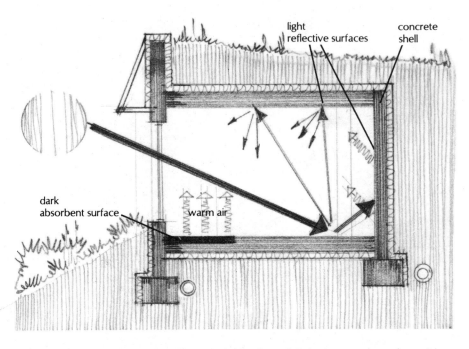

light
reflective surfaces

concrete
shell

dark
absorbent surface

warm air

*Figure 3–9: Surface color deter-
mines whether energy will be
evenly distributed around the
room or concentrated at a
surface.*

because the sun goes down. The maximum heat load occurs when there is no solar gain, so it becomes next to impossible to meet 100 percent of a load with direct solar, even with greatly increased mass.

You can use this graph as a basic estimating design tool. Make some photocopies of the graph in figure 3–8 so you can experiment with different window areas for the various south-facing rooms in your house. I must stress, too, that this is an exercise aimed at developing guidelines, and it should not be looked at as a strongly limiting factor in the ultimate design of your home. Properly used, this method will help to keep you from designing in too much direct-gain window area, which is the most common error in passive solar design. Oversizing a direct-gain system can actually result in *higher* annual utility bills because the oversized window causes greater nighttime heat loss, increased summertime cooling loads and possibly some overheating during sunny-but-not-too-cold winter days. Basically, the end result of the oversized direct-gain system is discomfort (not to mention money that is poorly spent).

The basic underlying assumption in creating this guide for direct-gain design is that room temperatures below about 65°F and above about 80°F are to be avoided. For some situations, that restriction can be lifted. A bedroom, for example, can rise to 90°F or even more during the day without creating a problem, since typically no one is occupying the space. By allowing higher daytime temperatures to occur, more heat can get into storage (remember that one of the factors that increases heat storage is a higher input temperature),

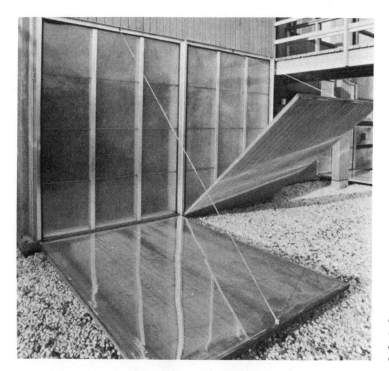

Photo 3-2: Exterior movable insulation for windows is the best barrier against winter heat loss and summer heat gain. The reflective surfaces in these panels increase solar gain in winter.

resulting in higher temperatures at night, when the room is occupied. So, a room that is essentially unoccupied during the day can have larger-than-guideline direct-gain windows without compromising comfort.

A second underlying assumption used in the guideline is that if overheating occurs, the excess heat must be bled off to the outside in order to maintain comfort. If there is a way to move the overheated air to other, cooler, parts of the house, then what would normally be considered an oversized direct-gain system becomes reasonable and starts to have the characteristics of isolated gain, another type of passive system.

What about room color? Because the room surfaces are the solar collector, the color of those surfaces has a lot to do with how the room reacts to the presence or absence of sunlight. A common misconception is that all room surfaces should be dark in color. Actually, the only room surface that needs to be dark is the surface of the primary storage. As mentioned eariler, for the primary storage to be effective, the conversion from sunlight to heat has to take place on the surface. Dark colors are much better for that conversion than light ones. On the other hand, surfaces not involved in storage, i.e., those which are of low mass or are covered with an insulating material, are best colored white or with light shades.

One of the most unpleasant characteristics of direct-gain systems is glare.

*Photo 3-3: These self-storing insulating shutter panels pro-
vide marked reduction in heat loss through windows while
solving the problem of storage when they're not in use.*

For the system to work, the sun has to get into the room. If room surfaces are
dark, great contrast exists between the interior of the room and the bright sunlit
view to the outside. That extreme contrast causes visual discomfort. Light colors
in the room help spread available light around, thereby reducing contrast.
Furthermore, spreading light around the room is also spreading solar energy
around the room, which increases chances of its absorption at secondary storage
surfaces. That effectively increases the efficiency of secondary storage. If it
appears that there will be enough light bounced to secondary storage surfaces to
take advantage of this effect (usually the case if floor coverings are light in color),
then the secondary storage surface can use a darker color than if little energy
reflection takes place. Figure 3-9 shows a typical room utilizing these concepts.

Improving System Performance

Several things can be done to reduce heat losses through windows at night.
The easiest is the addition of movable insulation in the form of shutters, panels,

Photo 3–4: Movable insulation panels that fit tightly into window frames work well and are inexpensive to make.

shades and drapes. This insulation should be tight enough along the sides and bottom to prevent cold air that forms on the glass surface from flowing into the room. It can increase the window R-value from about R-1.8 to as much as R-9 and significantly improve the performance of direct-gain systems. The use of movable insulation on nonsouth windows will also reduce the total room heat load. For those periods of the year when too much sun comes in the window, the insulation can also be used to cut down the effective window size. Notice that

Photo 3-5 (right): *This insulating quilted shade has five layers: two layers of fiberfill on each side of reflective plastic and two exterior quilted layers. The shade runs on side tracks, which create a tight seal.*

Photo 3-6 (above): *Thermal drapes that have top, side and bottom seals combine insulating capability and conventional appearance.*

when the heat load calculation method in Appendix 1 is used, movable insulation can be a factor in estimating the Home Heating Index.

Condensation on the glass is a problem with interior-mounted movable insulation, but insulation that has a vapor barrier and that fits tightly around the window minimizes condensation. Since there is some water vapor present in the air trapped between the insulation and the glass, condensation cannot be eliminated unless the insulation is either outside the window or in actual contact with the glass. Interior movable insulation such as foam panels with plastic film or foil coverings are easy to mount in contact with glass. To secure such panels,

Zomeworks Corporation in Albuquerque, New Mexico, manufactures Nightwall, a mounting system in which magnets are recessed into the insulation and metal strips adhered to the glass.

Summer Operation

For those times when the rejection of solar energy is more important than collection, several options are available for the direct-gain system. The first, using movable insulation, has been mentioned. That does, however, eliminate the window as a ventilator and as a source for view and daylighting. Also, if the movable insulation is inside, as is usually the case, it is less effective than exterior shading systems, which prevent the sunlight from even getting to the window. One exterior shading option is the roof overhang. The overhang length is calculated to take advantage of the fact that the sun is much higher in the sky in summer than in winter. By providing overhangs designed to be just long enough to block the summer sun, the winter sun, which is much lower in the sky, can shine in unimpeded. Figure 3-10 shows how these overhangs work and illustrates a method for calculating their length. The appropriate overhang length changes with latitude. Overhangs in southern states are going to be shorter than overhangs in the North. In the South, the solar altitude is so high in summer that little direct solar energy gets through a window, even without overhangs. On the other hand, the summer sun in Canada is low enough to be a real problem without overhangs or other window-shading devices.

Another method for shading direct-gain systems in the summer without seriously degrading their winter operation is to plant deciduous vines or trees to the south. Their leaves provide summer shade but conveniently disappear during the cold months. Movable shading devices such as awnings, exterior louvers and removable reflective surfaces also help provide relief from the summer sun.

Indirect-Gain Systems

The importance of coupling incoming solar energy to storage mass in passive systems cannot be overemphasized. The most difficult problem associated with implementing a direct-gain system is providing enough mass and mass surface area to allow the solar heat to be stored at a rate fast enough to prevent overheating. In an underground house, it is usually possible to have plenty of mass available. Unfortunately, most of it is in the form of secondary, not primary, storage.

If there were a way to provide a square foot of primary storage for every square foot of south-facing solar glazing, then most of the design problems of direct-gain systems would disappear. One way is to leave the concrete floor clear of carpets, rugs and most furniture. For some rooms, that is a practical solution, but for most it is not. Another way is to use a concrete exterior wall as the storage mass and to put glazing between it and the sun. By definition it is no longer a

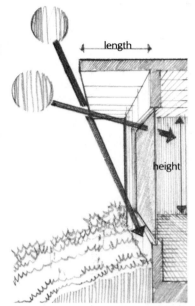

Figure 3-10: The following equation provides a quick method for determining the projection of a fixed overhang.

$$Projection = \frac{window\ opening\ (ht.)}{F}$$

where: F = factor from following table:

NORTH LATITUDE	F FACTOR
28°	5.6–11.1
32°	4.0–6.3
36°	3.0–4.5
40°	2.5–3.4
44°	2.0–2.7
48°	1.7–2.2
52°	1.5–1.8
56°	1.3–1.5

Select a factor according to your latitude. The higher values will provide 100 percent shading at noon on June 21, the lower values until August 1.

Source: Redrawn from Edward Mazria, The Passive Solar Energy Book, copyright © 1979, with the permission of Rodale Press, Emmaus, Pa.

direct-gain system. This new approach is called *indirect gain,* since the solar gain takes place outside the actual living space, though it is still a part of that room.

Designing for Indirect Gain

With the masonry surface placed immediately behind the south-facing glazing, not only does the storage mass always receive a maximum of possible sunlight, but the room does not have to cope with large amounts of energy coming in during the day, when it is least needed. As shown in figure 3–11, an indirect-gain system absorbs the solar energy on the outer surface of a masonry wall and delivers it to the room only after passing it through the wall's thickness. If the wall were made of a highly conductive material such as aluminum, the heat coming in from the sun would be distributed throughout the entire mass of the metal almost as fast as it could come in from the sun. So, for a high-conductivity wall, the temperature on the inner surface is hottest when the sun is brightest. A masonry wall, however, is of moderate to low conductivity. Concrete, in the thicknesses used for walls, distributes heat energy much more slowly than the sun is able to deliver it. That has two major effects on the way the system performs. First, because the energy can't get out of the way fast enough, the absorbing

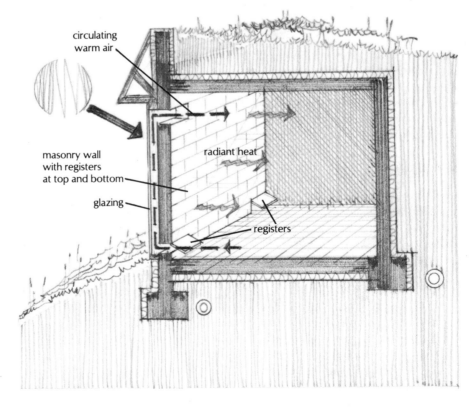

Figure 3–11: A Trombe wall type of indirect-gain system uses inlet and outlet vents for bringing solar heated air into the living space for daytime heating. Nighttime heating comes from heat stored in the masonry wall.

circulating warm air

masonry wall with registers at top and bottom

glazing

radiant heat

registers

surface has an energy ''pileup,'' which forces the surface temperature to go up. Second, the inner surface of the wall doesn't react to the temperature change on the absorbing surface for a substantial amount of time. This time delay is on the order of hours. A typical 8-inch concrete wall will delay entry of solar energy long enough to make its energy delivery after the sun sets. Since the greatest need for heat is after the sun goes down, this time delay is an advantage. The high exterior-surface temperature is not, however. The higher the temperature of the wall surface that faces the solar glazing, the greater the losses back to the outside. Remember that heat transfers more and more rapidly as the temperature differential between two surfaces goes up. A surface with a higher temperature will lose heat more rapidly than one that's cooler.

Now that we know the way the wall reacts to the presence or absence of solar energy, that knowledge can be translated into a prediction of how the unit will interact with the room behind. Of course, one obvious characteristic of an indirect-gain system is that it appears as a solid wall from the inside. Where a direct-gain system is characterized by an excess of sunlight with attendant glare, indirect-gain systems block sunlight from the living space. Where direct gain delivers heat exactly in step with the changing solar intensity, indirect gain has a built-in time delay of several hours. The lower operating temperature of a properly sized direct-gain system allows it to operate at a higher solar conversion efficiency than indirect gain, in which the outer surface is often above 100°F for much of the day. Also, from a construction point of view, the indirect-gain unit has both an interior wall and a window, while the direct-gain approach requires just the window, so costs are going to be higher for indirect-gain passive approaches. The advantage of having the heat delayed is a significant one, but the problems of higher cost and lower efficiency need to be overcome. While no one has figured out a way to avoid the cost increase, there is an effective method of improving operating efficiency. Remember that the reason for the lowered efficiency is the high temperature on the absorber surface. If we can in some way cool it down without losing the heat from the room, efficiency will improve. By circulating room air over the absorber surface, the absorber is cooled and the room is warmed during the day. System efficiency is then raised. We also get a bonus in the process. For walls of reasonable thickness (up to about 15 inches of concrete), the heat stored during the day is not enough to provide heat for more than one cold night, so the wall falls to below comfort temperature by the time the new day begins. Since it takes hours to get the collected heat to the inside, the wall is too cool in the morning. By circulating room air over the hot side of the wall — which warms up as soon as the sun's rays strike it — some of that chill is offset.

Figure 3–12 shows the classic indirect-system design. The drawing is from a patent application by Edward S. Morse in 1881. You may also know this design by its popular name of Trombe wall. In this approach, the glazing is spaced some 4 to 6 inches from the mass wall. Air can pass through openings in the wall at its base

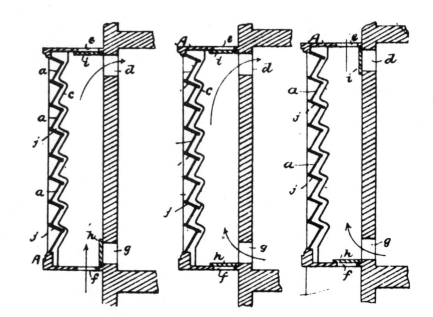

Figure 3-12: This drawing, from a patent application made by Edward S. Morse in 1881, shows three ways of operating this solar heater. Today we commonly call this device a Trombe wall.

and top. The air that is heated between the glazing and the wall rises and passes into the room through the top register, while cool air is drawn in at the bottom, creating a *thermosiphon* heat transfer system. Studies have shown that about 25 percent of the heat delivered to the room comes from the thermosiphon loop, and the remaining 75 percent is delivered through the wall as radiant energy — the most comfortable and effective form of heating.

Note that it is important to be able to shut off at least the lower convector register in the wall at night. If that is not done, the cold air that forms between the glass and the wall will fall to the bottom of the cavity and flow through the register along the floor of the room, causing drafts and discomfort. Although you can buy automated registers, their cost is very high.

Creating an indirect-gain passive solar heater for underground houses is straightforward. By simply glazing the south-facing structural (masonry) walls and providing a thermosiphon path between the wall and the room, the system is created. The cost of adding the glazing to the wall is usually less than the cost of providing exterior insulation and not much more than interior insulation.

Implementing Indirect-Gain Systems

Because the absorbing surface is immediately behind the glazing, it is very important that the surface be a good absorber of solar energy. To be effective, the color must be dark. Flat black is best, but deep red, blue, green, brown and purple are also effective colors, and they're more decorative. Experimenters have coated the exterior surface with materials that have light-absorption but poor heat-radiation characteristics (so-called *selective surfaces*) and have found that such walls have significantly higher efficiencies. But, selective surfaces for indirect-gain mass walls are not yet commercially available, and these systems are still very much in the experimental stages. Because of the inaccessibility of the absorbing surface, the dark coloration must be as permanent as possible. Material that is dark in color is best. A second choice is material treated with a penetrating stain, and a poor third is material that has been painted.

For the glazing, any material that is suitable for solar collectors is suitable for indirect-gain systems. Be sure that the manufacturer guarantees the material to have a long life at temperatures of around 150°F. Some plastic glazings will deform or gradually darken when exposed to sunlight for long periods at elevated temperatures. From a performance and appearance point of view, glass is hard to beat. Low-reflectance, etched-surface glass is excellent, although not required. Because such glass is not transparent, the seemingly strange spectacle of glass over a blank wall does not occur. Flaws in the concrete are also hidden, so the outside of the wall does not have to be finished. Even with clear glass, the dark surface of the wall often makes the completed unit look like a window opening into a darkened room. The minimum spacing of the glass from the wall should be about 4 inches. With less than that, there is too little room for an effective thermosiphon loop to be generated (when the space is vented to the room behind). Even if it isn't vented, too small a spacing causes increased losses through conduction to the cold glazing. Too great a distance between the wall and the glazing reduces the effective primary storage area due to shading of the upper parts of the wall. Large spacings also may require unusual construction methods, while 4- to 6-inch spacing can be achieved by hanging the glass on the wall.

What about cleaning the inside of the glazing? In most cases, the accumulation of dirt on the inside of the glass is not enough to be a problem in the operation of the system. It may be an appearance problem, however, if clear glazing is used. Where you are concerned about keeping the inside of the glass clean, either make the glazing panels removable from the outside (a bit of a pain, I assure you), or space the glazing far enough away from the wall to provide access for cleaning tools.

Because the interior surface must transmit the stored solar energy to the room by radiation, it cannot be covered with materials that might insulate the mass. Large paintings, draperies, paneling or furred-out drywall finishes are *verboten*. The color of the interior surface is not critical, but there must be no barrier to the flow of the heat from the interior of the wall to the visible surface. Perhaps the best finish is plaster applied directly to the concrete. A poor second choice is gypsum board nailed into intimate contact with the concrete, and an even poorer third choice is gypsum board that is glued to the wall.

Combining Indirect and Direct Gain

Solar houses have successfully used only indirect-gain techniques, but the combination of direct gain and indirect gain in the same installation has a lot going for it. The direct-gain portion provides quick warm-up, illumination and a view,

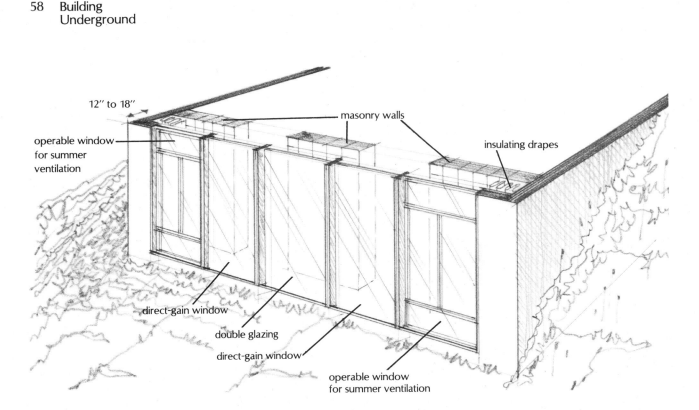

12" to 18"

operable window
for summer
ventilation

masonry walls

insulating drapes

direct-gain window

double glazing

direct-gain window

operable window
for summer ventilation

Figure 3–13: This high-performance solar heating system, as used in the Missouri Sun-Way homes, combines direct gain, indirect gain and movable insulation. For additional information on this solar concept, write to the Missouri Department of Natural Resources, P.O. Box 176, Jefferson City, MO 65102.

while the indirect-gain portion gives off a steady heat in the evening and reduces the probability of overheating from the direct-gain part. So, the combination provides fairly even heat to a room that is well illuminated, has a view and needs little special treatment on its walls, ceiling and floor. In sizing such systems, the direct-gain part is the most critical. As in direct-gain systems alone, a south window can be up to 15 percent of the room's floor area without particular concern for storage. That size allows for good lighting and view, plus a reasonable amount of heat gain to a well-insulated space. Sizing the indirect-gain system is not particularly critical for most climates, as long as there are wall thicknesses of about 8 inches. The thinner the wall, the higher the delivery temperature and the more quickly the heat is lost, so uncomfortable conditions can result from thin-walled systems in mild climates. Those installations that use 8-inch or 10-inch thicknesses perform quite well and are probably the most cost-effective.

A point of diminishing returns is reached in the area of indirect-gain systems. As a rule of thumb, it is reasonable to expect an indirect-gain unit to deliver from 300 to 600 Btu per square foot of glazing per day. Most rooms in underground houses will receive little benefit from systems larger than about 50 percent of the room's floor area. Houses that are in mild climates or are exceptionally tight and well insulated will probably benefit little from systems larger than 35 percent of each associated room's floor area.

Improving Performance

As with direct-gain systems, provision should be made to shade the system during summer months. Notice that in the case of indirect-gain units, the wall is the part that is most important. It is preferred that the glass be shaded also, but the wall is the absorber and must be shaded to prevent excess heating of the room. Some designs allow for venting of the inner cavity to keep heat from building up and being transmitted through the wall. Even more effective is a system that forces night air between the wall and the glass. This cools the mass, and about the time the sun is at its hottest, the surface is at its coolest. If the wall is poorly shaded, forced daytime circulation in the glazing/wall space may be useful.

Movable insulation can be used effectively in indirect-gain systems. The best place for that insulation is outside or between the glass and the wall. After the heat is stored in the wall, insulating the wall from the cold outside temperatures makes a big improvement in thermal efficiency. It is particularly effective when several days of cloudy weather occur. This insulation can transform a wall that's good for 24 hours of heat storage into one with 48 or more hours of storage capacity. The use of that insulation also relieves the occupant of the need to close the thermosiphon registers, as there is no longer cold air accumulating in the glazing/wall space at night. Unfortunately, automatic, electric-powered insulation systems have to be used unless the space is readily accessible. Such systems are available but expensive. Exterior insulation that fits snugly over the outside of the glass is excellent but is likely to be expensive and hard to install, and it usually can't be operated from the inside. Probably the best choice is to space the glazing far enough from the wall to allow manually drawn insulating drapes or shades to be installed. Figure 3–13 shows a particularly attractive design for this feature. It combines direct and indirect gain, with the insulation placed to take care of losses from both systems. Although movable insulation will improve the performance of any passive solar system, it may not be cost-effective in milder climates.

Isolated-Gain Systems

Both the direct- and indirect-gain approaches are integrated into the rooms they heat. A passive solar system in which the energy is collected separately from the room is an *isolated-gain system*. The most familiar example of an isolated-gain system is an attached solar greenhouse. Actually, any room dedicated to collecting solar energy to be delivered elsewhere is an example of isolated gain. Many are the names for such rooms: sun porch, solarium, sunroom, solar greenhouse, and sunspace. While solar greenhouse is the most common term used, that conjures up a vision of a plant-filled room complete with barrels of water for heat storage and a cat (a solar cat, of course) basking on the brick floor. That is perhaps too detailed a vision. By no means must a useful isolated-gain room have either plants or integral storage, although the cat will be there every sunny day. For our purposes, let's call the room a sunspace.

Some System Design Guidelines

For a room to be a sunspace, only two requirements need to be fulfilled. There must be south glazing, and the space must be physically separate from, but thermally attached to, the room to be heated. By thermally attached, I mean there should be a means for circulating heated air from the sunspace to other rooms. Solar heat may or may not be stored in the sunspace itself, and the walls of the sunspace may or may not take part in the heat transfer to the main house. Sunspaces are often free-floating thermally. That is, they usually don't have a back-up heating system when it gets cold, nor is there a need to be concerned about overheating on sunny days. But when sunspaces are often occupied, a back-up heating system may be needed.

From the point of view of the designer, isolated-gain systems allow the collection of solar heat from distribution to rooms having little or no southern exposure. Also, because the heat is collected outside the occupied space, distribution can be easily controlled. Automatic controls are much less costly for isolated gain than for either direct-gain or indirect-gain methods. From an energy point of view, sizing of the system is not critical to owner comfort, since it is isolated from the living space. The owner of a house with an isolated-gain system will pay more for this solar component than he would for a direct- or indirect-gain system, but the extra cost provides a usable room as well as a solar collector. All in all, the economics are quite good, considering the increased resale value of the house with its additional solar floor area.

The glazing requirements of an isolated-gain system depend on the orientation of the system. If the unit faces south, only the south wall needs to be glazed unless skylights or roof glazing is required to provide additional light for plants. In this case, there can be glazing on parts of the east wall, west wall and/or the roof. If the system faces southeast, the southwest wall should be glazed and the northeast wall solid. If the unit faces southwest, the southeast wall should be glazed.

To include storage or not to include storage? That is the question. For every Btu stored within the sunspace, a Btu is not available for delivery to the house. Also, internal storage lowers the air temperatures in the sunspace. To transfer the most heat to the house, high air temperatures are desirable, since the solar-heated air is to be mixed with a much larger volume of cool house air. On the other hand, the presence of some storage within the sunspace will reduce temperature swings in the space and will reduce heat losses from the adjacent house wall. To decide how much storage to include, several factors must be considered. The most important is how the sunspace will be used. If it is just a solar collector with only very occasional occupancy at sporadic times, then the tendency should be toward minimal storage. If the space is to be used to grow thermally fragile plants in the middle of winter, then lots of thermal storage is needed.

The makeup of the wall separating the house and the sunspace is a determinant as well. If that wall is well insulated, then heat losses through that wall

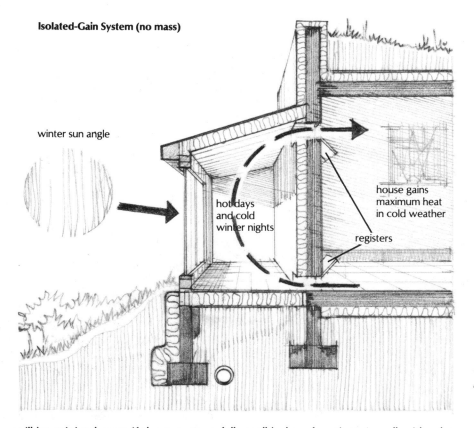

Isolated-Gain System (no mass)

winter sun angle

hot days
and cold
winter nights

house gains
maximum heat
in cold weather

registers

Figure 3–14: With minimal mass present, isolated-gain systems provide the maximum in warm air when the sun is shining.

will be minimal even if the sunspace falls well below freezing. A wall with a low R-value, such as one containing a lot of glass, will lose a lot of house heat back into the sunspace at night. Storage in the sunspace may reduce that heat loss when considered over the entire heating season. I stipulate the entire heating season because, theoretically, for a day in which all the solar energy that's received goes to heating the house, there should be little difference in the heat delivered, whether it is delivered directly or stored for later release. Over the season, however, there will be many days when the solar energy that's received exceeds the needs of the house for heat. Without storage, the excess heat is lost back to the environment. With storage, it is made available for nighttime use.

To put this all together into design rules, there are three factors that should be considered in making the "to store or not to store" decision. First, if the sunspace temperatures are not to approach the nighttime outside temperatures, then storage is needed. The more closely temperatures need to be controlled in the sunspace, the more storage is needed. Second, if low-R-value walls separate the sunspace from the house, increased storage is appropriate. And, finally, for large systems that often have surplus energy available, storage is useful to gain the full energy benefit that is possible from the system.

*Figure 3–15: With a floor or wall
surface of dark-colored concrete,
enough mass is available to keep
the isolated-gain space cooler in
the daytime and warmer at night
with only a moderate reduction in
heat available to the house.*

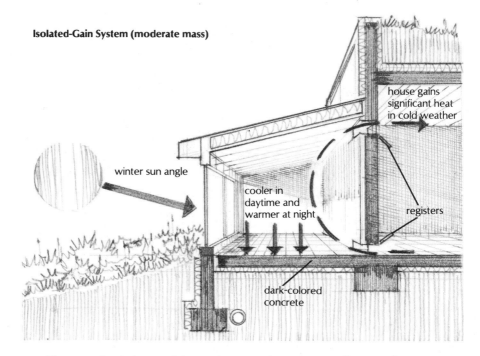

Isolated-Gain System (moderate mass)

winter sun angle

house gains
significant heat
in cold weather

cooler in
daytime and
warmer at night

registers

dark-colored
concrete

The most basic form of thermal storage is a concrete floor. In the sunspace,
carpeting and furniture are often not present, so the floor becomes available for
storage. In underground housing, the wall separating the sunspace from the
house is often structural concrete. That wall and the floor provide medium
amounts of thermal storage. From there on, the addition of internal storage
elements such as brick planters or drums of water is necessary.

With that in mind, let's design a sunspace that will normally not be occupied
and that will not be used to grow temperature-sensitive plants. For most units,
the size is almost always determined by considerations other than solar. Aesthetic
considerations, the different uses of the sunspace, the cost and the available area
are much more likely to determine the size than is the amount of energy needed
from the unit. Thus, it can be said that the desire for a sunspace over another type
of system shouldn't be primarily based on a desire for heat.

For our design, assume a 1,500-square-foot house with 60 linear feet of
southern exposure (not including the garage). With an 8-foot eave, we can have
a sunspace with 480 square feet of south-facing glazing that will span the entire
length of the house. That represents 32 percent of the floor area of the house.
With glazing representing up to about 25 percent of the floor area, storage for a
sunspace is not a critical matter. With more than that, storage is needed. So let's
add an insulated concrete floor. Further, much of the south wall of the house is
taken up by windows and two patio doors, which indicates a low average
R-value. Thus, more storage will be good. The concrete wall of the house will be

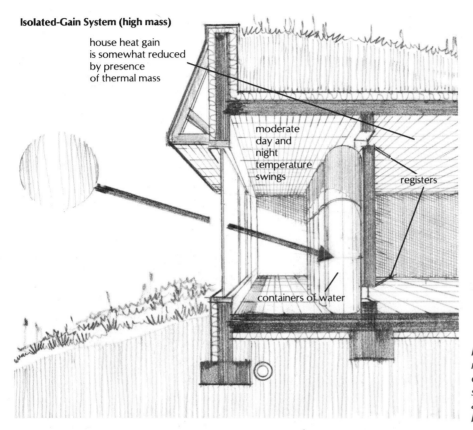

Isolated-Gain System (high mass)

house heat gain
is somewhat reduced
by presence
of thermal mass

moderate
day and
night
temperature
swings

registers

containers of water

Figure 3-16: You can use thermal mass in an isolated-gain space to control both the temperature swings in the solar space and the amount of daytime heat received by the living space.

a good storage component if it is not insulated from the sunspace. Because we need to reject as much heat as possible in the warm months (while losing the minimum in the winter), the end walls are built solid with plenty of insulation. So is the roof, which includes an overhang of appropriate length to shade the glazing in the summer. Inasmuch as there are times when overheating will occur because the house doesn't require the energy, a cross-ventilation system is needed. There are registers at opposite ends of the sunspace, with one near the ceiling at the end downwind of the prevailing summer wind direction and the other near the floor at the upwind end. The higher (outlet) register is downwind so that the ventilation will be aided both by the rising of heated air and the presence of any wind. The unit should be double glazed in colder climates and single glazed in warmer ones. The crossover is somewhere around 4,000 annual heating degree-days.

If the same house is planned to have a much smaller sunspace, for example, one that is 16 feet long, the design will be identical to the full-length unit except that there will be no need for storage, since the house will need all the heat the sunspace can provide in cold weather.

Photo 3–7: Solar greenhouses, or isolated-gain spaces, provide solar gain as well as additional living space.

Attached Solar Greenhouses

Designs for plant-growing rooms combined with isolated-gain systems are more complex because of the need for fairly close environmental control of light, temperature and humidity. A relatively large amount of storage mass is needed to reduce temperature fluctuations. A sloped south wall or a partially glazed roof will be needed to provide optimum light throughout the interior of the greenhouse. Construction materials must be able to survive in a high-humidity environment where lots of condensation is present.

The interior should be light in color except for those surfaces where storage takes place. Since it is important to have illumination well distributed in the greenhouse, large black or very dark surfaces are not appropriate. Rather than being concentrated in one area, storage should be well distributed so that dark and light surfaces alternate. Also, it is always best to keep storage masses in locations where they are in direct sunlight. End walls can remain solid, since east and west glazing are net losers of energy and contribute little to light availability. Also, if the end walls are transparent, considerable light will pass on through the greenhouse rather than being trapped inside where it can be converted to heat.

Controlling the Greenhouse Environment

It is a rare greenhouse design in a rare climate that maintains adequate levels of temperature, light and humidity without regular user intervention. When temperatures above those needed for growth occur, the surplus energy can be piped into the house if needed, or vented to the outside if not. When temperatures fall below those allowable for the plants in the greenhouse, heat must be added from the house or from an internal heater. Insulating covers for the glazing are a must in cold climates with intermittent sun. The interface between the greenhouse and the house must allow controlled passage of heat, but a filter should be present to keep insects from passing through also, particularly if you release ladybugs or praying mantises for biological control of pests. Inlet ventilation should be directed up toward the ceiling and not across the plants. Finally, artificial lights may be useful to extend photoperiods for small areas of the greenhouse during the shortest days of winter.

The usefulness of a solar greenhouse as a heat source will vary dramatically with the climate and with the way the greenhouse is managed. In cold climates that have a high percentage of available sunshine, such as the high country of Arizona and New Mexico, enough energy will be available to heat an area of the house comparable to the area of the greenhouse. Night insulation may not be necessary (although it will be a benefit), since the storage mass will be recharged on a daily basis most of the time. In cold climates that have many multi-day periods without direct sunshine, such as the Northeast, the solar greenhouse will be doing very well to hold its own during the coldest months, even with night

Figure 3-17: Four methods for moving heat from an isolated-gain space to the house.

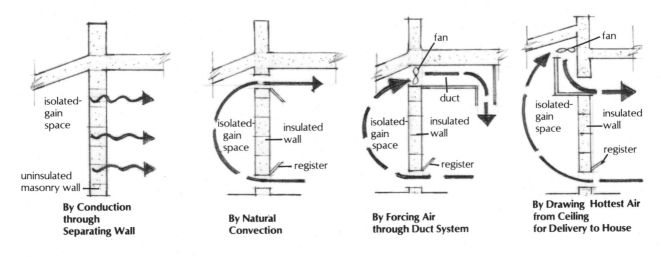

By Conduction through Separating Wall

By Natural Convection

By Forcing Air through Duct System

By Drawing Hottest Air from Ceiling for Delivery to House

insulation. During those periods, the only thermal benefit to the adjacent house will be reduced heat loss through the common wall. In the spring and fall, however, excess heat will usually be available, and due to reduced demand for space heat, the solar greenhouse will generally do a good job as an isolated-gain unit. What is often done in these more difficult climates is to use the greenhouse only as a season extender. By shutting down the unit as a plant grower during the coldest months (when growth, and particularly fruiting, may be limited by low light levels), removing much of the internal storage mass and using the room as an ordinary isolated-gain unit, more heat becomes available for home heating. Then, toward the end of the cold period, some storage is reinstated (water barrels refilled), and part of the greenhouse is used as a starting area for the next season's crops. By spring, all the storage should be back in place, and all greenhouse systems are "go" for another nine or ten months.

A unique option for underground houses is the roof greenhouse. It should be built directly on the concrete of the roof (which has been designed to accept the additional load) with skylights protruding up into the greenhouse space here and there to provide light to the rooms below. Ducts are installed to deliver fan-forced heated air from the greenhouse to the house, or the reverse, if need be. The unit can be pointed directly south no matter what the orientation of the supporting house, thereby offering a solar assist for "impossible" sites.

Figure 3–18: A roof-mounted isolated-gain space can be oriented to the south even if the house cannot. It provides winter heat, summer ventilation and daylighting.

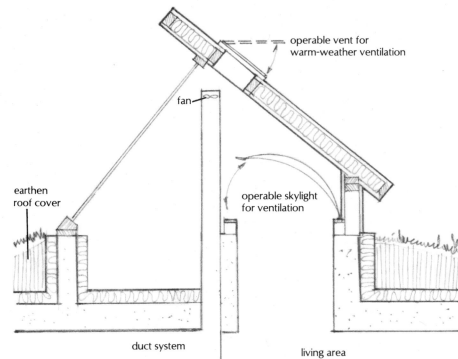

operable vent for
warm-weather ventilation

fan

earthen
roof cover

operable skylight
for ventilation

duct system

living area

Getting the Heat to Where You Are

In those cases where the sunspace is separated from living spaces by a single wall, solar heat from the isolated-gain room can sometimes be delivered through the natural convection that occurs when the heated air rises to the ceiling of the room. Although a doorway or a window passes some heat into the house, to make full use of the heat that has been collected, a couple of rules need to be followed. Since the warm air gathers at the highest point, heat passages (outlets) at ceiling height are best. That means that most windows — which stop a foot or more from the ceiling — won't deliver all the available heat. Special ceiling-height windows or through-the-wall heat registers need to be installed for best performance. The size of the opening is also important. Because there is little pressure behind the air movement, relatively large openings are needed. I have found that about 4 square feet of ceiling-level opening for every 15 linear feet of wall separating the sunspace from the house is a workable rule of thumb. There must be some method of opening and closing the vents. Crank-type casement windows installed with their tops at ceiling height are particularly nice, since the crank (mounted on the bottom of the window) is easy to reach. Other approaches include openings with sliding or hinged insulated panels, multiple louver windows, commercial registers with built-in dampers that close tightly, basement-type (awning) windows and even cabinet doors complete with magnetic latches. Whatever system is used, it should be easily operated with latches or controls at a comfortable height. Since it doesn't make any difference whether the shuttering device is in the house or the sunspace, let convenience and decor considerations be your guide.

Just as important as the path for the heated air is the location of a path for air to return to the sunspace. There must be a complete loop. By far the best way to create a thermosiphon, or natural-convection, loop is to allow the cool air at floor level to pass from the house into the sunspace. What we are really creating in our natural-convection system is a very low-power heat engine. It is a basic principle of heat-engine design that maximum efficiency is gained when the temperature difference between the input and output of the engine is greatest. For our heat engine, the maximum temperature differential is found between the floor and the ceiling, so to get maximum efficiency, return, or inlet, air should be provided at floor level, where the room air is coolest. A door is excellent. If for some reason, such as security or noise, you don't always want to have the door open during sunny days, cool-air registers (inlets) of comparable size and design to the warm-air registers (outlets) can be used. Remember that the amount of air circulated is pretty much limited by the smallest opening in the whole system, so cool-air and warm-air registers should be comparable in size for best heat delivery.

The horizontal placement of the air registers is not crucial, but in general, best performance will be gained when the placement of the inlet register is shifted away from the outlet. That causes the maximum ''sweeping'' of heat from the sunspace. One approach I like is to have a door (the cool-air return) centrally located, with the warm-air registers at the ceiling on either end of the sunspace. That approach provides good circulation at modest cost. Of course, the best (and most expensive) design provides inlet and outlet air registers along the entire length of the floor and ceiling. When that is practical, it is the preferred system.

In underground houses, it is not always possible to have an isolated-gain space located just across a wall from the main living area of the house, but there are other ways to use isolated gain to advantage. For example, most of the disadvantages of skylights can be overcome by placing an isolated-gain room over them. This aboveground extension of the underground house not only provides light through the skylight without the usual problems of leaks and heat loss, but it also gives you a greenhouse right next to your roof garden. Obviously, such an installation will not naturally convect heat to the house, so a forced-air system has to be provided. As in the naturally convecting installation, the heat should be gathered at the ceiling of the sunspace and the cool-air return placed to pick up the coolest air in the house proper. Deciding how big to make the fan is a problem. If the fan is too small, much of the solar heat will never get into the house. If it is too large, the temperature of the solar air will be little higher than that of the house air, and there will be the additional annoyance of noise and drafts.

A reasonable rule of thumb is to have the fan large enough to change the air in the sunspace from 10 to 15 times each hour. That may be a little slow during the middle of the day and a little fast in the early and late hours, but overall it is a reasonable average. If the sunspace is 8 feet high, 10 feet long and 6 feet deep, it has a volume of $8 \times 10 \times 6 = 480$ cubic feet. The fan must move at least $480 \times 10 = 4,800$ cubic feet per hour (cfh) or $4,800 \div 60 = 80$ cubic feet per minute (cfm), or at most $480 \times 15 = 7,200$ cfh or $7,200 \div 60 = 120$ cfm. Note, that is *actual* delivery. Most fans, however, particularly small ones, will deliver considerably less than their rated value when connected to a ducting system. Also, the fans have to work against the natural tendency of the warm air to rise. So, for short duct runs (2 to 10 feet), double the calculated value to get a rated value for the fan. For longer ducts (10 to 20 feet or so), triple it. Thus, if we have a 480-cubic-foot roof-mounted greenhouse, and the warm air is drawn in at a point 8 feet above the roof and delivered into the house from there, then a fan with a rating of around 150 to 160 cfm would be reasonable.

It would be even better to install a variable-speed fan and adjust its flow until the temperature of the air coming from the duct is between 75 and 90°F. If the air is warmer, increase the speed of the fan. If it is cooler, decrease it. An outlet temperature of about 85°F is really optimal. That is warm enough so that the delivered air does not feel cold as it moves over your skin. An alternative to using a variable-speed fan is to install an oversized unit and adjust the airflow with a damper. That has the disadvantage of wasting some electrical energy, but in fact, the motor on even a larger fan will draw little power anyway, and the wasted power is actually converted to heat, which is passed on to the house.

Sometimes fan systems are used even where convectors are practical. In the case where security is a prime consideration, the presence of relatively large through-the-wall openings may not be desirable, and the much smaller openings required for fans is preferred. Also, fans can be automatically controlled with thermostatic switches. In cases where the fan just blows through a wall and not through a long duct, the rule of thumb for fan size would be to choose the fan with a rating closest — but above — the design value. So in the above example, the design value was 80 cfm, and if your fan catalog lists 75-cfm, 100-cfm and 150-cfm units, choose the 100-cfm unit. Don't forget to include a way to return the air at floor level.

Automatic control is easy to implement in fan-driven systems. The thermostat turns the fan on when it senses a sunspace temperature of about 95°F, and it shuts the fan off at around 80°F, but no lower than 75°F. Most snap-action-type thermoswitches with set-points in this range are satisfactory. Sensitive thermostats, such as those used to control whole-house heating systems, can be used, but they are usually low-voltage, low-current devices that require a power relay in the circuit to turn the fan on and off; in short, they're more complicated and expensive than the task requires. The temperature sensor, which may be a remote sensor or may be enclosed in the thermostat, should be in open air (not in contact with a building surface) at the same level as the fan intake and preferably not more than 3 feet away.

A variable-speed fan may be operated by a *proportional controller,* which increases the fan speed with higher sunspace temperatures and reduces it to off as the sunspace temperature falls below about 80°F. These controllers are

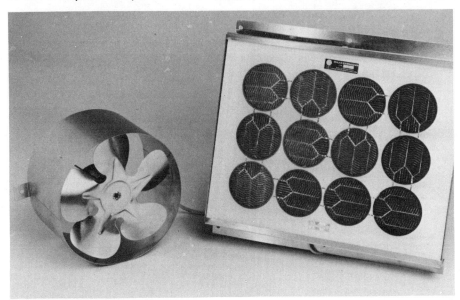

Photo 3–8: A fan powered by photovoltaic cells is usually adequate to draw heat from an isolated-gain space into the living area. In most installations, such a fan eliminates the need for extra wiring and a controller.

designed to control ventilating fans or air conditioners, and usually the controller and the fan must be purchased as a matched unit. *Multi-speed fans* that are not continuously variable are also available with step-type thermocontrollers. Such a unit should turn the fan on at its low speed with inlet temperatures of 75°F and raise the speed in steps until it is at full speed at 90°F and above. Variable or multi-speed fan systems provide better operating efficiency and a generally higher comfort level than single-speed units, but they have higher first-costs and sometimes greater maintenance requirements because of added complexity.

Manual control of a thermosiphon or natural-convection heat-delivery system is easy. The simple opening and closing of the registers is all it takes. The cool-air registers at the floor must be closed at night if the sunspace is unheated. Otherwise, cold night air ends up in the house along the floor. They may be left wide open during sunny days without fear of having excess heat flow when it is not needed. The cool-air registers need to be manually opened in the morning and closed at night, but the warm-air registers may be automatically opened and closed using any of several automatic greenhouse venting units currently on the market. Those are operated by a device that expands when heated and shrinks when cooled, and this expansion is converted to movement of an arm that is attached to a fixed base and to the vent panel. The opening and closing of the warm-air registers is automatic, depending on sunspace temperature. The operating temperatures for these units should be the same as for the electrically

Photo 3-9: An automatic greenhouse venting device can be used to open and close vents in isolated-gain spaces. This "heatmotor" operates hydraulically in relation to temperature, opening a vent when the inside air increases in temperature and closing the vent when the air cools.

powered devices. They should be shut at below 75°F and fully open at 90°F and above. With these devices, remember to open and close the inlet registers manually. Lightweight plastic flaps that are forced closed by cold air falling and opened by warm air rising are used in some installations. While they automatically open and close as desired, all the units I have seen either have a life of only one season or less or seriously restrict the air flow. Until someone can provide a very lightweight film that can withstand the thermal and mechanical stresses involved, I recommend using rigid dampers.

Solar Skylights, Atria and Monitors

Conventional horizontally mounted skylights are *not* effective providers of solar heat, except in the summer. The sun's rays reach skylights at too flat an angle in the winter to provide much heating, while in the summer the high sun angle allows direct penetration into the house. These units, even double-glazed ones, are heat losers par excellence, and there is always Wade's Observation: *Skylights leak.* I don't consider conventional skylights a useful addition to underground houses. If you disagree, at least do yourself a favor by providing insulated covers for the units to keep the house heat in when it is cold and the sun's heat out when it is hot.

But what about the need for daylighting? What options other than skylights can provide light for northern rooms? My previous admonition was for *horizontally* mounted skylights, but if the units are mounted vertically, or nearly so, they become solar collectors in the winter and useful lighting (but not heating) devices in the summer. There are several ways to achieve this goal. The first is to mount the skylight unit itself horizontally but to install a reflector over it, constructed to reflect low-angle sunlight down into the skylight while shading the skylight from high-angle rays. In fact, the reflector can be curved to achieve some solar concentration in the winter months. This approach does not improve the heat-loss characteristics, however. To do that, the reflecting apparatus has to be enclosed with a layer of glazing on the south side, and the rest of the enclosure must be insulated. The end result is a *monitor skylight.* Such a monitor reduces heat losses. Summer heat gains are greatly reduced, but daylighting and winter heat gains are provided. Leaks are less likely because the monitor structure provides an additional barrier against water entry if proper waterproofing procedures are followed.

The use of an atrium with a glazed roof is sometimes included in underground designs to improve solar utilization. Unfortunately, glazed atria suffer the same problems as skylights but on a larger scale. They add significant amounts of heat in the winter. They also add a lot of heat in the summer. Domed, gable-type and flat atria (not suitable in snow country) all create grossly overheated conditions in the summer and reasonable but not spectacular heat gain in the winter. Considering the cost of such installations, I cannot recommend glazed atria in general. If they are included in the design, it is imperative that excellent ventilating capabilities are included. With most glazed-atrium designs, it is easy to provide hot-air exits but not so easy to figure out a way to get the outside air into the system at ground level. Drawing ventilating air into the atrium through house doors or windows will result in drawing hot outside air into the house except in climates where summer air temperatures are rarely in the 90s. That is not desirable, since one of the main advantages of an underground house is that it can stay significantly below ambient temperature due to its contact with the earth and its great mass. About the only really satisfactory way of providing outside air input to an atrium is by providing a special duct that runs to the outside and delivers its air at floor level. With very few exceptions, then, glazing an atrium is not very satisfactory unless the glazing can be easily removed for the summer or can be externally shaded.

Photo 3–10: Solar domestic hot water systems are cost-effective solar applications, but they may not be especially attractive. These solar collectors, mounted on the garage roof of the Pearcey house (see chapter 8), are hidden from view by a low wall.

Solar Water-Heating Systems

The economic picture for active solar water heating is considerably better than for active space heating. Domestic hot water (DHW) systems have a constant, year-round load, and, in colder climates at least, passive water-heating systems may be summer-only operations, while properly designed and installed active DHW units work fine all year. My preference for the location of DHW

Photo 3–11: If protected against freezing, a passive solar breadbox can preheat water effectively year-round. In some climates, a breadbox can provide all of the domestic hot water in summer.

solar connectors is anywhere but over the house itself. The reason for this is a simple one, which follows one of Wade's Laws: *Don't put anything through the roof unless you have to.* Leaks, cracks and unknown horrors cluster around roof penetrations, and in my experience, it is much better to have 20 extra feet of well-insulated hot-water-pipe runs than to have shorter runs and continuing problems with the roof. Other than that, there is no particular difference between active DHW installations for underground houses and similar ones for conventional housing.

Passive DHW systems are either thermosiphon systems, in which there is a separate collector and tank, or breadbox-type systems, in which the tank is mounted in an insulated box with a glazed south side. Thermosiphon systems require the tank to be higher than the collector for good performance, which is generally not easy to provide in underground installations. Also, in freezing

Photo 3-12: *Zomeworks' Big Fin collectors, mounted to the structure of this isolated-gain space, provide inexpensive hot water on sunny days.*

climates, collector freezing is a serious problem, and high-reliability provisions must be made to avoid damage from ice formation in the collector. Breadbox units perform poorly in cold weather unless the glazing is covered at night with good insulation and the unit is attached to the house so that there are no connecting pipes (even insulated ones) exposed to freezing conditions. The breadbox unit costs the least of the solar DHW systems but is the poorest performer. An active system costs the most but is, when properly installed, the best performer. The difference in performance is greatest in the coldest climates, so active systems for DHW look good in the middle and northern states and Canada, while passive systems are a good choice for the Sun Belt.

DHW systems can also be installed inside an isolated-gain room. A breadbox heater or a system using Zomeworks' Big Fin collectors can be included inside the room for hot water. If that is done, don't forget that the solar energy that goes into hot water is no longer available for space heating. To overcome that loss, a larger amount of sunspace glazing can be provided.

Summary of Passive Systems
Direct-Gain Systems

Direct-gain passive solar systems are the simplest and least expensive of the solar options. On the negative side, direct-gain systems have wide temperature swings because storage is usually poorly coupled to incoming energy flows during the day, and the low R-value of the window allows greater heat losses at night than in other solar designs. Glare can be a problem if rooms having direct-gain systems are dark in color, particularly on the floor and ceiling. Although many direct-gain solar houses have performed well, direct gain is best used to supplement other passive or active solar systems and in those spaces where the glare and temperature-swing characteristics do not cause problems.

Indirect-Gain Systems

An indirect-gain system is created by glazing south-facing masonry walls. For underground houses, rarely more than 50 percent of a room's floor area needs to be installed as indirect-gain solar for cost-effective applications. The solar wall should be from 6 to 10 inches thick. Coupling the room air to the air between the glass and the wall through a series of floor-level and ceiling-level registers improves system efficiency as well as reducing the chilling effect of the cold wall in the morning.

Glazing-to-wall spacing should not be less than about 4 inches nor more than is reasonable to allow full sunlight to fall on the maximum wall area in the winter. Usually, spacings over 18 inches should be avoided. Combining direct and indirect gain provides the best of both worlds by mutual compensation for the failings of each. Reasonable sizes for a combination would be 15 percent of the room's floor area in direct gain and 35 to 50 percent in indirect gain for a 5,000-heating-degree-day climate. Movable insulation can dramatically improve performance in cold climates when installed either outside the glazing or between the glazing and the wall. In milder climates (4,000 annual heating degree-days or less), movable insulation for indirect-gain systems may not be cost-effective. Although interior color is not important to the performance of the indirect-gain system, the interior surface of the mass should not be decoupled from the wall itself. Any material applied to the concrete should have a thermal conductivity about the same or better than concrete, and it should have total contact with the wall.

Isolated-Gain Systems

Unlike either direct- or indirect-gain passive solar systems, there is no practical upper limit to the size of an isolated-gain system other than

(continued on next page)

Summary of Passive Systems—*Continued*

economics and the available space at the site. The flow of collected solar heat is easily controlled and surplus heat can be ignored, dumped to the outside or transferred to storage without affecting the comfort level of the house. Although its cost is greater than that of other passive approaches, the added floor area is valuable and should be considered in the economics of the application. An isolated-gain unit facing south requires glazing on only the south wall unless additional light is needed for plants. The requirement for storage within the sunspace is dependent on the intended use of the space and the properties of any wall separating the sunspace from the house. If the space is to be used as living space at night, storage, at least in the form of an exposed masonry or concrete floor, is to be desired. If the space is to be used for growing plants in the winter, storage should be maximized. If the space is to be used only as a solar heater, then minimal storage is appropriate. If the common wall is concrete or masonry, no additional storage is needed unless winter vegetables are desired. If the common wall is poorly insulated, with lots of glass, for example, some storage is desirable. If the wall is well insulated, little storage is needed. Both thermosiphon and forced-air systems are used for moving the solar heat into the house. The thermosiphon systems have the advantage of requiring no electrical hookup but do require fairly large openings between the house and the sunspace. Automatic control is easiest with forced-air designs, and wall penetrations can be smaller. Fan sizing is important to ensure maximum heat transfer without drafts. Movable insulation for the sunspace glazing is important only if maintaining comfortable nighttime temperature in the sunspace is important. As a general rule, the more important sunspace storage mass is, the more important is night insulation. Thus, night insulation is more important if the space is used for growing plants than if it is just a daytime solar heater. Obviously, in colder climates, more night insulation is economically beneficial.

CHAPTER 4

The First 15 Feet

Although there are many reasons why people need houses, certainly one of the most important is environmental control. For comfort, people need to have their surroundings kept within a fairly narrow range of temperature and humidity. Rain is nice for the garden, but most of us prefer to keep it out of the living room. The sun's heat is welcome when it's cold outside, but not when it's hot. A shelter, be it a skyscraper or an underground dwelling, has as its primary purpose the isolation of its occupants from the outside environment. Its walls have to be constructed in a manner to effect that isolation in as cost-effective a manner as possible, so the outside environment always must be considered in the design process. We are all familiar with the aboveground environment, since that is where we spend most of our lives. But the underground regions are known only to miners, underground builders or those who work in or study the subsurface environment, and even they rarely know much about many of the conditions found in this netherworld. The designer/builder of an underground house does need to be familiar with the climate below the surface and the character of the medium (earth) surrounding the house.

Understanding Soil

I suspect that if we were a race of burrowing creatures, we would have a great many words to replace the single term, soil, just as the Eskimo has many words for snow and ice. The material that we so casually call soil has such a diversity of character that samples from any given site are almost unique, but there is enough commonality in soil characteristics to allow classification by such factors as its chemical type, its particle size and its reaction to mechanical stress and heat flow.

Chemical Classification

Since all of the inorganic content of a particular soil comes from some parent-rock material, the composition of that rock can be used for classification. The simplest is *residual soil,* that is, soil that has not been moved from the site of its formation. This soil is an extension of the underlying bedrock, with particle size and organic content the only significantly variable factors. Residual soils are usually fairly shallow; a depth of 10 feet is a lot for such a soil. Since they are usually shallow and have a uniform character, most residual soils provide a readily predictable construction environment. The very shallowness of residual soils means that underground construction is not too easy, however.

Most sites that have the necessary depth for fully underground construction have soils that have been moved considerable distances from the parent rock. Transport by glaciers, water and wind often results in soils being deposited thousands of miles from where they were formed from rock. The chemical composition of the parent rock is still apparent, of course, and much of the variation in soil types depends on this chemical composition. For the agronomist or the gold panner, chemical composition may be the most important characteristic of a soil, but to the builder, particle size is usually more important.

Classification by Particle Size

Table 4-1 shows a common system of classification based on the size of soil particles. Mechanical characteristics, such as load-bearing ability, shear strength, dimensional changes when wet, and cohesiveness, depend mainly on the size of soil particles, the range of particle sizes present in a sample, and the water content.

Obviously, water content is a variable for most soils, so to simplify matters, completely dry — in fact, oven-dried — soil is used as the basis for particle-size classification. You are probably already familiar with several size classes of soils. Gravel is a class of soil (many people don't think of gravel as a soil, but for engineering purposes it is) with primarily large particles. Sand is a soil with mostly small particles. In general, the smaller the particles, the more dry soil acts like a liquid. If you stand on dry sand, you can expect to sink into the soil surface as sand is pushed up alongside your feet. If you stand on coarse gravel, there's very little horizontal movement of the soil under your feet, and there's very little sinking. When the soil-particle size is microscopic — essentially rock flour — its resemblance to a liquid is striking.

The thing to notice in particular is that vertical pressure on the soil results in movement of the soil not only down (under your foot), but horizontally (from under your foot) and up (alongside your foot). Your weight forces soil particles to be pushed out from under your foot, so the sinking is not just due to compression of the soil (see figure 4-1). The ability of a soil to resist this movement in directions other than that of the applied force is called *shear strength*.

Photo 4-2: Soil horizons are obvious in this excavation for an underground house.

If the soil sample is confined so that it cannot move horizontally, then placing a weight on the soil will still result in a little sinking. Since the rock particles themselves are not compressible — at least not by any force commonly found in construction — the movement is due to a tighter packing of the soil particles. The pore size is reduced as the particles are redistributed into a smaller volume of space. If the weight is released, the dry soil does not expand back to its former size, since nothing was actually compressed. All that happened was that the soil was more tightly packed. This process is called *compaction,* and the force with which a soil resists a compacting force is called its *compression strength*.

It is very important to notice that the compression strength varies with the level of compaction of a soil. If a soil has been compacted to the maximum extent, that is, if the particles are squashed together tightly, then compression

Table 4-1
Soil Components and Fractions

SOIL	SOIL COMPONENT	SYMBOL	GRAIN SIZE RANGE AND DESCRIPTION	SIGNIFICANT PROPERTIES
Fine-Grained Components	Silt	*M*	Particles smaller than 75 μm sieve identified by behavior: that is, slightly or nonplastic regardless of moisture and exhibits little or no strength when air dried	Silt is inherently unstable, particularly when moisture is increased, with a tendency to become quick when saturated. It is relatively impervious, difficult to compact, highly susceptible to frost heave, easily erodible and subject to piping and boiling. Bulky grains reduce compressibility; flaky grains, i.e., mica, diatoms, increase compressibility, produce an "elastic" silt.
	Clay	*C*	Particles smaller than 75 μm sieve identified by behavior; that is, it can be made to exhibit plastic properties within a certain range of moisture and exhibits considerable strength when air dried	The distinguishing characteristic of clay is cohesion or cohesive strength, which increases with decrease in moisture. The permeability of clay is very low; it is difficult to compact when wet and impossible to drain by ordinary means, when compacted is resistant to erosion and piping, is not susceptible to frost heave, is subject to expansion and shrinkage with changes in moisture. The properties are influenced not only by the size and shape (flat, platelike particles) but also by their mineral composition; i.e., the type of clay-mineral, and chemical environment or base exchange capacity. In general, the montmorillonite clay mineral has greatest, illite and kaolinite the least, adverse effect on the properties.
	Organic Matter	*O*	Organic matter in various sizes and stages of decomposition	Organic matter present even in moderate amounts increases the compressibility and reduces the stability of the fine-grained components. It may decay causing voids or by chemical alteration change the properties of a soil, hence organic soils are not desirable for engineering uses.

SOURCE: From T. William Lambe and Robert V. Whitman, *Soil Mechanics,* copyright © 1979, reprinted with the permission of John Wiley & Sons, New York, N.Y. (Adapted from A. A. Wagner, "The Use of the Unified Soil Classification Systems by the Bureau of Reclamation," *Proceedings of the Fourth International Conference on Soil Mechanics and Foundation Engineering,* London, England, 1957.)

NOTE: The symbols and fractions were developed for the Unified Classification System. The sand fractions are not equal divisions on a logarithmic plot; the 1.7 mm was selected because of the significance attached to that size by some investigators. The 380 μm size was chosen because the "Atterberg limits" tests are performed on the fraction of soil finer than the 380 μm.

SOIL	SOIL COMPONENT	SYMBOL	GRAIN SIZE RANGE AND DESCRIPTION	SIGNIFICANT PROPERTIES
Coarse-Grained Components	Boulder	None	Rounded to angular, bulky, hard, rock particle, average diameter more than 300 mm*	Boulders and cobbles are very stable components, used for fills and ballast and to stabilize slopes (riprap). Because of size and weight, their occurrence in natural deposits tends to improve the stability of foundations. Angularity of particles increases stability.
	Cobble	None	Rounded to angular, bulky, hard, rock particle, average diameter smaller than 300 mm but larger than 150 mm	
	Gravel	G	Rounded to angular, bulky, hard, rock particle, passing 75 μm† sieve retained on 4 mm sieve	Gravel and sand have essentially same engineering properties, differing mainly in degree. The 4 mm sieve is arbitrary division and does not correspond to significant change in properties. They are easy to compact, little affected by moisture, not subject to frost action. Gravels are generally more perviously stable, resistant to erosion and piping than are sands. The well-graded sands and gravels are generally less pervious and more stable than those which are poorly graded (uniform gradation). Irregularity of particles increases the stability slightly. Finer, uniform sand approaches the characteristics of silt: i.e., decrease in permeability and reduction in stability with increase in moisture.
	Coarse		75 to 19 mm	
	Fine		19 to 4 mm	
	Sand	S	Rounded to angular, bulky, hard, rock particle, passing 4 mm sieve retained on 75 μm sieve	
	Coarse		4 mm to 1.7 mm sieves	
	Medium		1.7 mm to 380 μm sieves	
	Fine		380 μm to 75 μm sieves	

AUTHOR'S NOTES:
*mm stands for millimeter
†μm stands for micrometer

strength is at the maximum. On the other hand, if compression strength is measured on a recently filled area, it will be much below the maximum. If you have ever had the pleasure of digging a hole in the wrong place and having to refill it with the soil you removed, you know that the soil you took out will not fit back in the hole! If a measurement of compression strength is made on the dirt used to refill the hole and on the dirt next to the hole (which has the same structure but has not been disturbed), the fill dirt will be much weaker in compression than the undisturbed soil. Therein lies a very important lesson for all types of construction: Place load-bearing foundations on undisturbed soil only. If construction on fill is the only way possible, the soil must be compacted either by letting nature do the work over a period of months (preferably years) through cycles of wet/dry and freeze/thaw, or by the use of special soil-compacting equipment.

Since compaction is simply the rearrangement of soil particles so that they fit more closely together, any process with that as its end result will work. Compacting equipment either places a high force on the soil surface, literally squeezing the particles into more intimate contact, or strongly vibrates the soil surface, causing the particles to shake into a closer arrangement. Some equipment does both. When mechanically compacting earth, two factors will greatly improve the chances of success. The first is the thickness of the layer being compacted — in most cases, it should be less than a foot, although 6 inches is better. Thin layers of earth allow forces that are applied on top of the fill to be transmitted throughout the thickness. If the fill is too thick, compacting forces on the surface will spread horizontally, and the bottom of the fill will not receive as much force as the top. The second factor is moisture. For every type of soil, there is an optimum amount of moisture needed for compacting. If the soil particles are very dry, they will rub against each other and resist compaction. A moderate amount of water will act as a lubricant and allow easy movement of soil particles into their new, close positions. Too much water, however, fills the voids between soil particles, and to get proper compaction, the water must be first squeezed out of the pores — a process that requires a lot of energy with fine soils.

Water Content of Soil

Soil can be characterized by chemistry or dry particle size, but the way soil reacts, both to mechanical stress and to heat-energy flow, is also strongly dependent on water content. I have already noted how the addition of water to a dry fill soil can reduce its compression strength through a lubricating effect. That effect arises because the water tends to adhere to the surface of the soil particles in a process called *adsorption.* Soil chemistry is the primary determinant of water adsorption. For example, soils with compositions derived from limestone will hold onto surface (adsorbed) water more strongly than will granite-based soils. This *adsorbed water* is bound fairly tightly to the soil particles and thus is very difficult to eliminate. To test for adsorbed water content, a soil is first air-dried and weighed; it is then oven-dried at temperatures higher than the boiling point of water and weighed again. The weight difference is the weight of the adsorbed water. If the sample is then left in the open air, it will gradually return to its original air-dried weight, indicating that the soil particles can literally suck moisture from the air.

Another way water is held in soil is through *capillary action.* The force that causes a water drop to form a sphere in free-fall also causes water to climb into small openings against the force of gravity. That force, surface tension, is such that the smaller the opening, the farther the water will move. This capillary "attraction" is often demonstrated in science classes by touching the end of the very small glass tube (called a capillary tube) to the surface of a liquid. The liquid suddenly rises several inches in the tube. Fine-grained soils also have very small

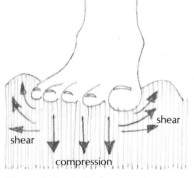

Figure 4-1: Vertical pressure produces both shear and compressive stress on soil.

Photo 4-3: Once soil is disturbed, compaction is necessary to force it back to its original state. According to the manufacturer, this upright rammer delivers 600 blows per minute, each with a force of 1,110 pounds.

spaces between grains, and water can rise many feet into such soil due to capillarity. Note, however, that for capillary action to happen, there has to be an air/water interface. A soil saturated with water does not exhibit capillarity. As water evaporates from the surface, creating an air/water boundary in the soil, water is brought up from saturated soils below through capillarity. For most soils, the water present in the upper levels consists of adsorbed water and water present through capillary action (*capillary water*), draw from deeper, wetter soils.

Because both adsorbed and capillary water are effectively trapped in the soil — adsorbed water by the surface attraction of the particles and capillary water by the surface tension of the water itself — they can be present in soil without creating waterproofing problems.

Waterproofing problems are caused by the third category of soil-borne water: *gravity water*. As you might surmise from the name, gravity water is water that is free to move through the soil in the direction of the force of gravity. Where gravity water occurs, the soil approaches or reaches the saturation point. The amount of gravity water that can flow from saturated to unsaturated soil is determined by the difference between the total water content and that bound up by capillarity and adsorption. Since both capillarity and adsorption are greatest in small-particle soils, gravity water will be a larger percentage of total water in a coarse saturated soil rather than in a fine saturated soil. Also, the larger the soil particles, the more easily the water can move, since there's less friction.

The measure of the freedom with which water can move through a soil is *permeability*. For example, many fine clays tie up water so tightly through capillarity and adsorption that there is no room for gravity water, and they are essentially impermeable. A particular clay called *bentonite* binds water so strongly that it is impermeable and is actually used as a waterproofing material (see chapter 6). At the opposite extreme of permeability are sandy and gravelly soils.

It is not uncommon for a layer of permeable soil to be bounded above and below by soil with poor permeability. Where that happens, the permeable soil acts as a conduit for groundwater, creating what is called an *aquifer*. If the layer of permeable soil slopes downward, considerable water pressure can develop. Artesian springs, which have enough natural pressure to shoot water into the air, are found in such formations. If the excavation for your underground home uncovers an aquifer, you will have continuing water problems. For drainage away from the house, you can build a short aquifer using high-permeability soil, usually gravel or sand, with lower-permeability soil on one side and the impermeable (we hope) wall of the house on the other. The water will take the easy way out through the aquifer instead of through the surface of the house.

The Water Table

Except in very thin soils or soils in very dry climates, there is usually a depth at which the soil is saturated with water. That depth is called the *water table*. If you dig a hole that penetrates the water table, water will come into the hole. It is then called a well (or a basement if you are unlucky). Above that depth, the soil may have considerable water content, but it is bound in place by capillary and adsorptive forces, so it does not pour out into the hole. Although the name water table implies an even, flat surface like a lake, rarely is the level of soil saturation actually a horizontal plane. Since water moves at varying speeds through different types of soil because of varying permeability, capillarity and adsorption, the rate of flow downward through the soil (called *percolation*) varies, and so do any horizontal flows. Thus, right after a rain, the water table may be close to the surface in an area of high permeability surrounded by lower-permeability soils.

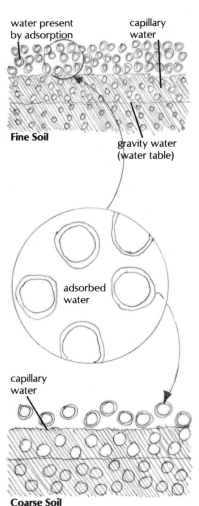

Fine Soil

water present by adsorption

capillary water

gravity water (water table)

adsorbed water

capillary water

Coarse Soil

Figure 4–2: Fine soil will hold much more adsorbed and capillary water than coarse soil.

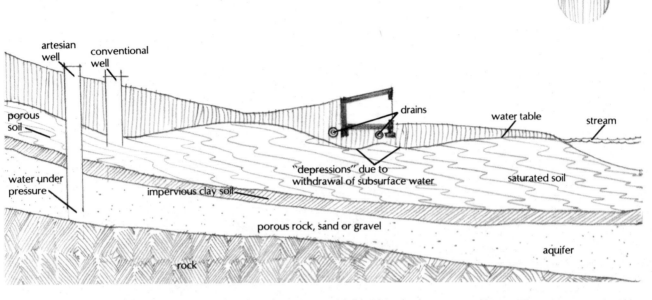

Also, aquifers may define the saturation level as they rise and fall within the layers of rock and soil surrounding them.

Soil Grading

Soil grading means the relative distribution of soil particles of different sizes in a given sample. If there is a small range of particle sizes, which means that all the soil particles are about the same size, then the sample is said to be poorly graded. If the sample contains a wide range of particle sizes, evenly distributed throughout, the soil is well graded. Residual soils and glacier-transported soils are usually well graded. Soils that were transported by wind or water, such as wind-deposited loess and the soils of river deltas, are often poorly graded. A fairly common situation is to find several fairly distinct soil layers, called soil *horizons*, some of which are poorly graded and some well graded. From the point of view of a person wanting to support a heavy building on a patch of soil, well-graded soil is almost always preferred. The presence of a wide range of particle sizes means that there is great opportunity for particles to be in physical contact with other particles over most of their surface. As shown in figure 4–4, poorly graded soil particles may contact other particles over only a small percentage of their surface. With large contact areas comes high friction, and with high friction comes high shear and compression strength. In construction, well-graded soil is often created by mixing several poorly graded soils in order to create a strong bed for a road or concrete slab.

Figure 4–3: There are many factors affecting the level of the ground-water table; important among them are different soil types, the presence of streams, lakes and wells, and the actual shape of the land, which can slow down or speed up rainwater runoff.

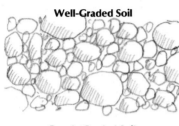

Well-Graded Soil

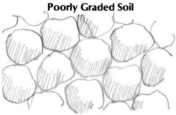

Poorly Graded Soil

Figure 4–4: The range of particle sizes and their distribution determines soil grading. Well-graded soil has little interstitial space, with a wide range of particle sizes evenly distributed in the soil. Poorly graded soil has large interstitial space, with a narrow range of particle-size distribution.

Organic Content

Another important variable in soil composition is the relative presence of organic materials. Vegetable and animal matter in varying stages of decay, and living organisms are found in widely varying quantities in soils. Some soils, such as peat, are almost all organic in content. Others, a desert sand for example, may be very low in organic materials. Generally, the upper part of a given soil — the so-called topsoil — has most of the organic content at a particular site. While organic material is necessary for life, high-organic-content soils are poor choices for construction sites due to their low strength and great variability.

Soil: Its Changing Character

The effect of changes in water content on the mechanical characteristics of the soil varies according to the composition of the soil. Some clay soils expand well over ten times in volume between the dry state and the saturated state. That expansion can exert great pressure on house walls and floors, sometimes causing cracking or even heaving. Other soils that are quite stable mechanically become very unstable with the addition of water. An example is the soil on many hills of Southern California. When dry, this soil supports itself and houses with no difficulty, but when sufficient moisture is added, it becomes very unstable, and the hillsides, complete with houses, slide to the bottom. Certain types of sand, when mixed with sufficient water, become "quick," the nemesis of many a jungle villain. Not to ignore opposites, some soils have greater strength when moist than when dry. Beach sand, when moist, readily supports cars, but when dry, good luck.

Site Drainage

A useful and common practice in underground construction is to place a layer of gravel completely around the home. This very coarse soil is little affected by varying water content, and by removing gravity water, it stabilizes the surrounding soil and isolates the house from variations in soil strength. Even if perfect waterproofing could be accomplished, good drainage would still be important.

Soil Density

Soil density is its weight per unit volume, usually per cubic foot or cubic yard. In a house that is covered with earth, the weight of that covering soil will be a function of its depth and its density. Very deep soil cover — a thickness at least equal to the width of the house — would actually be nearly self-supporting because the vertical force is distributed horizontally through soil-particle contact. But most underground-house roof cover is in the 2- or 3-foot range and has to be considered as dead weight. The soil density can be expected to change with water content, so, once again, the structural engineer will need to know what

Photo 4-4: Soil is a complex material. It not only varies from site to site, but its characteristics can change greatly at the same site. Distinctly different soil layers are called soil horizons.

kind of soil cover is to be placed on the roof in order to design the supporting structure properly. Few soils exceed 130 pounds per cubic foot, so the soil loading on a roof with a 3-foot cover will not exceed 390 pounds per square foot and will generally be more like 300.

The Underground Climate

Since one of the primary reasons people have for building underground is energy conservation, it is very important to understand the thermal conditions that exist down to a depth of at least 15 feet. While a great deal of information exists concerning temperature profiles at different depths and the thermal character of various soil types, almost all of it relates to essentially undisturbed soil and has been researched primarily in the fields of agriculture and geophysics. It is known that building a house underground can markedly modify the local underground climate near that building because the soil is disturbed and because of heat loss from the house. So far, only a few studies directly relate to a determination of the climate near an underground structure, but they have done much toward providing a basic understanding of the processes involved and some validation for theories about those processes. To better understand those ideas, some study of thermal processes in undisturbed soil is needed.

The Earth/Air Interface

The energy flows that significantly affect most soils occur at the surface. While in a few isolated places, geothermal energy strongly affects soil temperatures, over the great majority of the earth's land surface, the heat that comes from deep in the earth is insignificant when compared to the flows of energy at the junction of the air and the ground. The sun usually provides thousands of times more energy to the soil surface than ever comes from the earth's deep interior, so the first soil characteristic that has an effect on the underground climate is its surface character and orientation. Exposed dark-colored soils are warmer than light-colored earth. Likewise, south-facing slopes have higher temperatures than do flat surfaces or slopes facing other directions. Surface shading by vegetation has a considerable effect on soil temperature near the surface. For example, grassland soils are cooler in the summer than plowed ground.

Of course, other energy flows besides the solar input are important at the earth/air interface. While direct solar radiation leads the pack in adding energy, air convection adds heat when the air is warmer than the soil, as often happens in the spring. Warm rains can initially add considerable heat, while the "reverse" of rain, evaporation, is the leading cause of heat loss. Heat is lost also to cold air flows, cold rains and heat radiation to outer space. Any way you look at it, the earth's surface is a very dynamic place from an energy point of view. With all that going on at the surface, what is it like a few feet below the surface?

Figure 4–5: The sun provides most of the heat to soil, while surface evaporation and cold winds take heat away from soil. In desert climates where the air is dry and plant covering sparse, radiation may be the primary cause of heat loss. Geothermal heat significantly affects soil temperature in only a few places in the world.

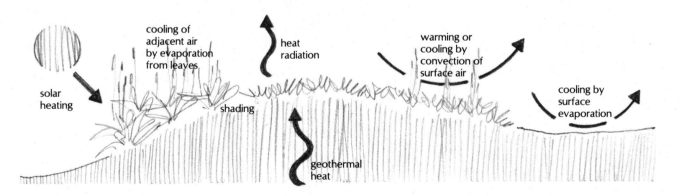

solar heating

cooling of adjacent air by evaporation from leaves

heat radiation

warming or cooling by convection of surface air

cooling by surface evaporation

shading

geothermal heat

Thermal Characteristics of Soil

We are used to rapid changes in weather. A cold front passes and thunder sounds. Rain falls and cold winds blow. The sky clears, warm air flows around our bodies, and a seemingly complete reversal of conditions occurs in just a few days. Rapid weather changes are the result of two primary characteristics of the atmosphere: its low heat capacity and the ease with which air convection occurs. Low heat capacity allows rapid heating and cooling to occur, while the ease of convection allows cold air to push warm air out of the way, and vice versa. If air could not easily move from one place to another, the weather would be a lot calmer, and if it took a lot of energy to change the temperature of air, then there would be practically no weather changes at all. That is exactly the situation within the earth. Outside of rare earthquakes and mud slides, masses of soil do not move from one place to another, so convection is nonexistent. With a mass of as much as 130 pounds per cubic foot, considerable heat has to flow into or out of that mass to change the temperature very much — orders of magnitude more than for an equivalent volume of air, since soil is 1,000 times denser than air. The end result is no weather (no change), only a stable climate. Rapid change is rare belowground, and any belowground change is very small in comparison to what happens aboveground.

The ideas of thermal conductivity and R-values take on new meanings when considered in relation to a continuous, three-dimensional mass like soil. In figuring heat loss through a wall, we assume a temperature difference for the inside and outside air and make calculations accordingly. The unmentioned assumption that is made when temperatures are assigned to the inside and the outside is that heat which flows through the wall will have no effect on the outside temperature. In the case of an aboveground house, that is generally true. Although right at the wall surface, heat is transmitted to a thin outside air layer, it is swept away and replaced by "new" air as rapidly as it forms. That is not the case in the underground climate. If heat flows through a wall into the ground, the ground changes temperature because the heat cannot rapidly get away without convection. In nature, the same sort of effect can be observed in earth/air heat transfer. At the very surface, soil temperatures change rapidly and in step with air-temperature changes. If the surrounding air changes 5°F in an hour, the top fraction of an inch of soil will also change about 5°F in an hour. The next fraction, however, may only change 4°F in that hour and may not start to change until 15 minutes after the air begins to change temperature. The smaller temperature change occurs because with increasing depth, more and more mass is available to absorb heat flows from the surface. The time delay occurs because it takes time for the mass above to change temperature, and the more mass, the more slowly the change takes place.

The same process applies to greater and greater depth. The effects of mass on the rate of heat flow are magnified, and increasingly large time delays occur

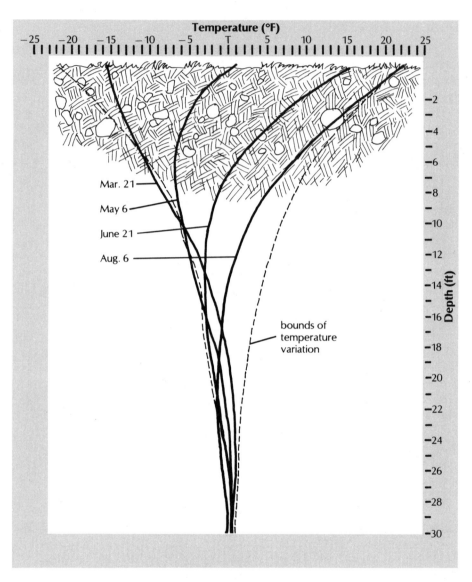

Figure 4–6: This graph shows typical temperature variation with depth for different times of the year. The T represents temperatures at constant temperature depths, which are the same as the temperatures plotted on the map in figure 4–9. Use as an example a site with a ground temperature of 58°F. On June 21, the temperature at a depth of 8 feet is 58 − 2 or 56°F. The temperature at a depth of 4 feet is 58 + 5 or 63°F.

SOURCE: Redrawn with the permission of Kenneth Labs, Undercurrent Design Research, New Haven, Conn.

between surface temperature changes and subsurface changes, with increasingly small variations relative to surface variations.

Observations show that for average soil (whatever *that* is) every foot below the surface introduces a one-*week* time delay and temperature-averaging effect. That means that variations that are noted at the 1-foot depth reflect not hourly nor daily but weekly average temperature changes, and those average changes show up a week later than the surface change that was the cause. Daily

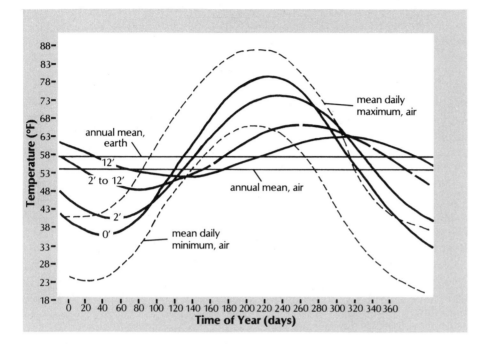

Figure 4-7: In this graph, temperatures at constant temperature depths are compared to mean daily and annual mean air temperatures. The temperatures represent generalized conditions for Lexington, Kentucky.

Source: Redrawn with the permission of Kenneth Labs, Undercurrent Design Research, New Haven, Conn.

temperature changes are often considerable, but when a week's changes are averaged, the change is much smaller. At 4 feet, the averaging effect has extended to a month. So, by 3 or 4 feet, the temperature variation is the result of seasonal change rather than the reflection of the passage of a cold or warm frontal system. Remember also that the amount of variation is lessening with depth also, so if the average temperature change over the month of September is 10°F, by the 4-foot level the change may only be 5 or 6°F and will show up at the end of October. At the 15-foot level — about the depth of the floor of an underground house — the time delay approaches six months, and the variation has been reduced by a factor of about four or five. So, if the difference between the May and August average temperatures is 20°F, that will cause a variation of only 4 or 5°F at 15 feet some four months later. If the relationship were exact, at 52 feet there should be no variation and the temperature should be precisely the average of the annual surface temperature. While that is surprisingly close to the truth, things are not quite so simple in the real world.

In order for the changes from surface to depth to be smooth curves, one basic assumption is involved: that the thermal characteristics of the soil are uniform throughout. As I am sure you realize by now, that is not likely to be the case. Soil composition can change several times, and quite markedly, between the surface and the 15-foot depth. Water content, in particular, is an important variable in thermal analysis, just as it is in mechanical analysis. With the same temperature change, wet soil can absorb more heat than dry soil, so the rate of heat conduction is changed. Also, although soil can't convect, water can. I have heard several theorists describe possible heat movement through soil at fairly rapid rates due to gravity-water flows. The rapid flow of surface water down through gravel fill on its way to the footing drains represents a rapid coupling of the house environment with the surface environment. Considerably colder house environments may have to be considered in the case of cold water rapidly percolating through gravel in contact with an underground house, such as might be the case in spring, when snow melts in northern climates. So far, this condition has not been observed in detail, but there are data to indicate that such flows should be considered in the insulation design.

Anything that changes the surface energy-flow patterns will also change those below. With the main energy-additive process being solar radiation, shading of the surface or the presence of a light surface color will reduce solar energy absorption and surface temperatures. On the heat-loss side of the ledger, evaporation is the primary process. Wet surface soils naturally have more evaporation losses than dry soils. Soils exposed to dry winds are cooled more through evaporation than are sheltered soils. Vegetation induces cooler surface temperatures due to transpiration of water through foliage and its attendant evaporation, as well as shading from the sun.

Figure 4-8: In climates with heavy snow, a "shell" of cold water may form around an underground house that has gravel or high porosity soil in contact with its surfaces. Separating the drainage path from the house may not only eliminate that possibility but can also make backfilling easier.

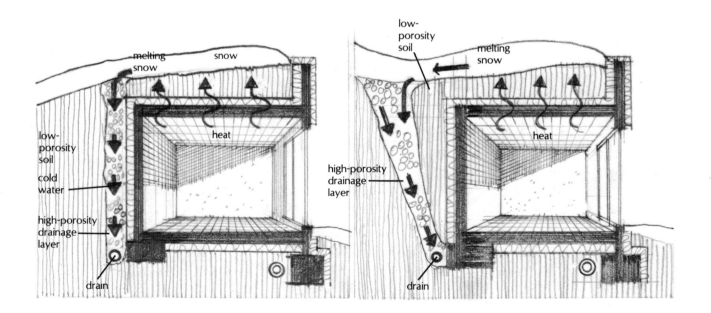

The relationships can be complex, however. Vegetation also reduces cooling by wind; wet soils change temperature more slowly and may stay warmer during the fall than dry soils; and snow reflects most of the sun's rays but insulates the ground surface against colder air above. Even the presence of shading in the summer seems to change only the time when maximum soil temperatures occur, at least below a thin soil layer at the surface. That happens because air temperature doesn't move in step with the sun, but maximum air temperature occurs 40 to 50 days later than the summer solstice. Likewise, minimum air temperatures occur 40 to 50 days after the winter solstice. Where the soil surface is not in the direct sunlight, the air temperature is the main influence on earth surface temperatures, so the sun's heat still influences the ground temperature, but indirectly, through the air, and with a six-week delay.

To summarize, increasing depth decreases temperature variation and increases the time delay of response to surface temperature changes. The primary variable of a soil's thermal character is its water content. Soil composition has an effect, but it is not nearly as important, except as it affects the soil's ability to hold water. Because the surface is the only significant place where the soil is in contact with a different energy environment, what is present at the surface of the soil has a great effect on how energy flows at the earth/air boundary.

Comparing the Surface Climate with the Subsurface Climate

With the increasing averaging effects, time delay and reduced temperature variation that come with increasing depth, it is reasonable to expect there to be some depth at which the temperature is a constant value that approximates the average annual surface temperature. That is the case, except that for most places, the constant temperature value is about 3°F warmer than the average annual surface temperature. That is usually attributed to the slight but measurable addition of geothermal heat from below. The depth at which constant temperature occurs depends on two factors. The first is what you consider to be constant temperature. If you are willing to consider a stability of plus or minus 5°F as constant temperature, then a constant temperature depth for the central United States will be around 10 feet down. If you insist on plus or minus 1°F, then you must go down to 25 or 30 feet in the same location.

The second factor in determining constant temperature depth is the difference in temperature between summer and winter. It is important because wide seasonal variations in temperature will show up deeper in the soil than will small seasonal temperature variations. For that reason, coastal regions, which are often characterized by relatively small differences between winter and summer average temperatures, will have a constant temperature depth closer to the surface than a site in the middle of the continent that has hot summers and cold

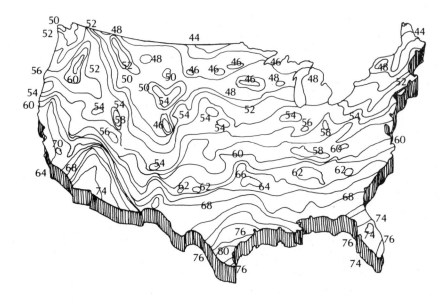

*Figure 4–9: Temperatures at con-
stant temperature depths.*

Source: Kenneth Labs, *Regional
Analysis of Ground and Aboveground Climate* (Oak Ridge,
Tenn.: Oak Ridge National Laboratory, 1981).

winters. Note, however, that once the constant temperature depth is reached,
the actual temperature may be the same for both locations. As the map in figure
4–9 shows, after constant temperature depth is reached, the temperature is
mainly a function of latitude and altitude — latitude because the distance from the
equator is the function that determines solar gains on a flat surface, and altitude
because surface temperatures fall about 4 to 5°F for every 1,000-foot increase in
height above sea level. Both effects are clearly seen on the map. The central part
of the continental United States is dominated by flatland where deep soil
temperatures change in step with the latitude. In the east in the Appalachian
Mountains and in the west throughout the high mountains, the contours are
jumbled and follow the contours of altitude rather than latitude.

Locally, there may be site-specific effects as well. Sites that are characterized
by soils containing a lot of water will have constant temperature depths nearer
the surface than those with drier soils. Valley bottoms will generally be colder
than hillsides or hilltops, because cold air flows down the hills and accumulates in
the valleys. Also, water is usually near the surface in valleys and evaporative
cooling at the ground surface may be higher.

The Climate around a Building

The atmosphere is such an efficient heat sink that a building aboveground
modifies its climate so minutely that it is of no design consequence. Underground,
that is no longer true. When an object of a different temperature than the

surrounding earth is placed in the underground environment, heat flows occur much like those flows coming from or to the surface. The relationship is complicated, however, by its occurrence in three dimensions. Essentially, the thermal environment generated from the surface energy interface is one dimensional. Unless there are unusual horizontal variations in soil characteristics, a probe moved straight down from any point on the surface in a radius of hundreds of feet or more will show the same temperature/depth relationship. Suppose, however, that someone buries a spherical container of radioisotopes 100 feet deep (see figure 4–10). Then assume that the isotopes give off a constant amount of heat, so that if a probe is sent along lines starting at the center of the container, temperature will change in some predictable fashion with increasing distance from the center of the container until the energy is damped out and undisturbed soil temperatures are regained. The effect is constant temperature shells around the box, which decrease in temperature with distance. The reason they decrease with distance is that as the heat moves away from the container, each increase in radius means that a considerable increase in soil volume is involved. For a 1-foot radius increase near the container, a mass increase of a few tons of soil may be involved. For a 1-foot radius some distance from the container, hundreds or thousands of tons will be added. In fact, since the volume of a sphere increases as the cube of the radius, doubling the radius will increase the mass in the sphere by eight times. The finite heat that the container can give off is very rapidly absorbed in the huge mass of the earth found only a few feet away.

Although this is a three-dimensional relationship, it still is not very complex, but neither is it very realistic. Picture again the same small heat-producing object, but this time at a depth of 10 feet instead of 100 (see figure 4–11). At this shallower depth, the temperature dampening effect is different in every direction because the undisturbed earth temperature is not a constant between the surface and the box below the container for at least 15 or 20 feet. In addition, the relationship changes with the seasons—not in step with them but delayed a little above the container, a bit longer alongside the container and a great deal below the container. In this environment, heat-flow patterns from the heat source are very complicated, indeed.

Now take one more step in your imagination and increase the size of the container from a small sphere to a volume the size and shape of a house (see figure 4–12). Instead of approximating a sphere, you are now working with a series of planes facing in different directions, each one except the floor and roof having a surface at depths from a couple of feet to 10 feet or more. All the complexities of the small heat source are still there plus weird things that happen at corners, uneven heat transfer from different parts of the surface of the house, the added presence of atypical soil in the form of added drainage gravel and fill hauled in from elsewhere, and probably a completely different surface cover than is present some distance from the house. The conclusion is that, unlike aboveground situations, an underground building extensively modifies its own

Figure 4–10: The heat given off by this object, which is buried 100 feet deep, will dissipate in predictable heat-flow patterns.

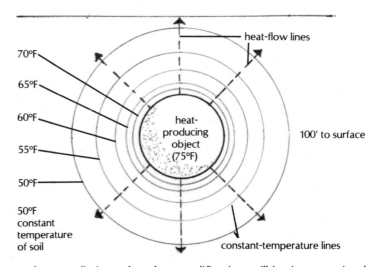

environment, but predicting what that modification will be is not a simple task. Yet, understanding what the thermal environment becomes is critical to energy-efficient, cost-effective design.

In an attempt to understand what is happening, several elaborate computer models have been created. While they are useful, thus far they are correct in only two dimensions and are poorly validated against the real world. A large data base of actual measurements needs to be created so that existing models can be validated or improved to better represent actual conditions.

To date, only a few underground houses have been studied for their thermal effect on the surrounding earth. Gordon Moore of the University of Missouri at Columbia, Missouri, did some pioneering work several years ago, and at the time of this writing, the University of Minnesota and Oklahoma State University have instrumented several underground house environments and are taking and analyzing data. Because of the interaction between climate, soil type and house design, many measurements at many different, well-documented sites will have to be made before any general conclusions can be made. A few things are clear already. The first is that it takes *months* for an underground environment around a house to stabilize. If you move into a new underground house in January, the winter will be over before you have reached thermal equilibrium with the earth around you. That means that for the first winter, fuel bills may be a lot higher than the next winter, when the earth surroundings are at equilibrium temperature. The amount of insulation around the house will determine how long it takes to reach equilibrium. This knowledge gives good justification for recommending different thicknesses of insulation at different depths and, when properly understood, results in rules for the best thicknesses and placement of insulation in different soils and climates, as is shown in chapter 6.

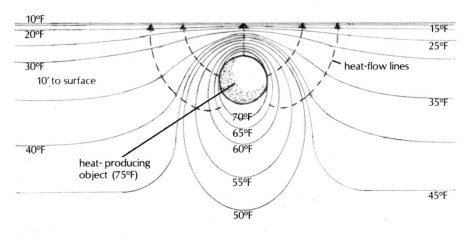

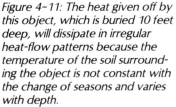

Figure 4-11: The heat given off by
this object, which is buried 10 feet
deep, will dissipate in irregular
heat-flow patterns because the
temperature of the soil surrounding the object is not constant with
the change of seasons and varies
with depth.

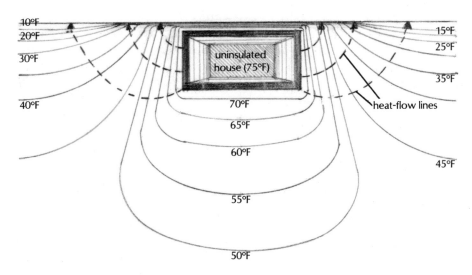

Figure 4-12: An underground
house can modify the temperature
of the soil surrounding it, but since
heat will not be dissipated evenly
from the various surfaces of the
house, it is difficult to determine
heat-flow patterns. This illustration
shows one possible pattern of
heat flow from an uninsulated underground house.

Designing for the Underground Environment

It would certainly be folly to build an aboveground house in Alaska that was designed for the mild climate of Hawaii. It would also be folly to build an underground house designed for the north-central plains, which has one type of soil and one very specific thermal character, at a site in southern Georgia, which has a completely different soil type and temperature environment. Even if the constant temperature depth and temperature were the same, the total underground environment, both physical and thermal, would be very different due to the different soil structure. To be successful in meeting structural, waterproofing and insulation needs, a house must be designed to fit the actual site.

Subsurface Site Analysis

Although subsurface conditions are usually quite constant at any given site, there is a great deal of variability in underground environments among different sites, and to have the greatest success, a careful analysis of the environment *at the actual building site* must be made. In particular, studying the composition of the soil and the way it reacts with water is important. At one site, a wall may need 50 percent less reinforcing than at another. A roof that may be clear-span at one site may have to have supporting columns at another. One site may require minimal waterproofing, while another may require a hermetically sealed polyethylene bag. And these sites, which have such different characteristics, may be only a few yards apart.

To be sure of the depth of the water table at a particular building site, it is necessary either to check it at that site or to get help from a soils engineer or hydrologist who knows how the water table lies at the site. Some people think that checking the depth of the water in wells around a site can indicate the depth of the water table, but unless an unused, *dug* well is within 50 feet of the proposed location, such data is not reliable. Driven wells or bored wells may extend to depths much greater than the water table in order to take advantage of deep aquifers with high rates of flow or to reach water with lower mineral content than water found near the surface. Near my home, there are dug wells less than 20 feet deep that have water all year, while many of my neighbors have driven wells with water no shallower than 100 feet. Unless groundwater levels are well defined at your site, drilling a test hole to a depth that is 5 or 10 feet beyond the bottom level of the house-to-be is a good idea. It is important to realize that the level of the water table varies with the quantity of rainfall in the recent past for almost all locations. Checking the level after a dry summer will yield lower values than those obtained following a wet winter. Also, the presence of drains and the proximity of producing wells and springs affect the depth of soil saturation according to the rates of water removal and the permeability of the soil. A well that is being used as a water supply will locally draw down the water table for many feet in all directions if the surrounding soil is reasonably permeable. Indeed, when drains are placed under a floor or around a foundation, they keep water away from the building by locally lowering the water table to the level of the drain tiles. This lowering effect extends furthest in layers of high-permeability soil, so if the soil is not very permeable, such as a clay soil, then

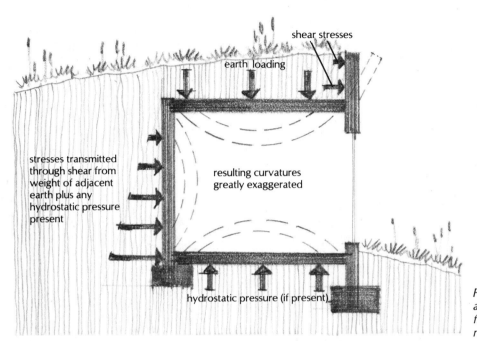

shear stresses

earth loading

stresses transmitted through shear from weight of adjacent earth plus any hydrostatic pressure present

resulting curvatures greatly exaggerated

hydrostatic pressure (if present)

Figure 4-13: The different forces acting on an underground house from the surrounding earth are significant design factors.

drain tiles may need to be spaced more closely than if the soil is of high permeability, such as sand.

The cohesiveness or shear strength of soil varies considerably with water content, and the way it varies is determined by particle size and particle composition. In general, a soils engineer will recommend design features so that an underground house can survive under the poorest conditions that are likely to occur. To do that, knowledge of the soil at and for several feet below the footing depth is required, and in most cases, a soil sample is necessary. Also, the structural engineer will be very concerned with what soil pressures may occur on building surfaces. Again, soil samples are important. A soils engineer can take these samples in the right place to determine the exact character of the soil. The soils engineer's generally modest fee, like the structural engineer's, is cheap insurance and almost always is saved in construction costs or repairs over the long run. The Yellow Pages and a listing from the state engineering license bureau or the local chapter of the American Society of Professional Engineers (ASPE) are good sources of names for soils or structural engineers. You should also seek referrals from builders and/or owners of underground houses or other engineers you know. Choosing an engineer who has had experience working with underground houses is usually preferable from the point of view of both fee and competency.

CHAPTER 5

Basic Construction Techniques: Concrete and Wood

Underground houses come in nearly as great a variety of forms as their aboveground counterparts. You can find underground houses made of rammed earth, log ends, wood frame, concrete block, steel, poured concrete and a multitude of combinations. There are even modular underground houses on the market. About the only type of dwelling not found underground is the mobile home! Concrete is the most common material used in underground construction, and there are many different construction systems possible using this versatile material. If concrete is your choice of building material, you still have to decide between sectional pours; monolithic pours; pretensioned, prestressed panels; posttensioned, prestressed construction; sprayed ferrocement shells; concrete blocks; or some combination of methods. This chapter looks in detail at each of these uses of concrete, with particular emphasis on special problems of underground construction. As you read, keep your site in mind and consider all the while how each of the different concrete building methods fits your needs.

Concrete

By far the majority of underground houses built to date are made of some form of concrete. Concrete offers many advantages over other materials. It is strong, available almost everywhere and essentially waterproof unless cracked or improperly installed. I suspect the greatest advantage is that it is the most familiar construction material for the underground parts of conventional housing, so there are a lot of contractors around who can work with it. But their being familiar with concrete construction only through house foundations, floors and basements can be a problem. There is a big difference between constructing a tolerable basement and constructing a good underground house, and the successful basement contractor may be hard to convince of that fact. The underground house is deeper in the earth, so stresses are greater. Waterproofing must be perfect in the underground house, while a modest amount of leakage during exceptionally wet weather is almost expected in a basement. Wall, floor and ceiling surfaces of an underground house need to be quite flat, which is not always the case with a basement wall. Any surface honeycombing, form seams (ridges that occur where form sections meet loosely) or other defects have to be removed if they're present, which is an expensive process that is best avoided by high-quality form work and proper working of the concrete. Finally, structural cracks (those that go all the way through) are unacceptable in the underground house but all too common in basements.

Photo 5-1: Concrete, by far the most common building material for underground houses, is also versatile and beautiful, as the house of Ethel and Bill Schlegel shows. The roof is made of precast, prestressed concrete panels, the walls of concrete blocks and the floor of poured concrete. The roof of the house is basically flat, although the front precast, prestressed panels above the sunspace and the south-facing window are tilted up to allow maximum solar gain during the winter and to retain the earthen roof cover. The tilted panels above the window also function as an overhang to block the summer sun.

Engineering Equals Insurance

How much would you pay for an insurance policy that would guarantee your concrete work against cracks, leaks and structural problems? A hundred dollars a year? Two hundred? You can buy such a policy for a one-time charge that is less than what you might be willing to pay for just one year's insurance. That "insurance" is the fee paid to a competent structural engineer for the design of your house so it won't crack, leak or structurally fail. Every site and every house design is different from an engineering standpoint, so even if you have a complete set of engineering drawings for your dream home that were engineered for a site other than the one you have finally chosen, do not use them. Get a reevaluation from an engineer for your new site. Chances are the new site will require some changes in the structural design that, if ignored, will result in a house that is either seriously overbuilt and needlessly costly up front, or underbuilt and needlessly costly later. While it is possible that a contractor experienced in building underground houses might be able to come up with a satisfactory "seat-of-the-pants" design, remember that it won't be his seat sitting in your house when problems develop. I strongly urge you to go to an engineer for the main structural design. In selecting an engineer, don't be afraid to compare résumés and prices. Surprisingly enough, you probably will get the best price from the most experienced engineer because it will take less time for him to provide you with a design. There are few structural engineers specializing in underground houses, but quite a few do a lot of small commercial concrete work, and the design criteria are similar.

Since you are hiring an engineer to do your concrete design, why should you worry about the techniques of concrete placement and structural engineering? The best design in the world is worthless if the building is not built according to plan. In most communities, the only concrete contractor available for underground construction using poured concrete will be the local basement builder. This competent tradesman has probably built hundreds of basements with few complaints and probably no structural failures. He will take one look at your engineer's plan and proceed to tell you how he can save you lots of money by leaving out all that unnecessary reinforcement and by loosening those overly stringent concrete specifications. He will correctly tell you how much experience he has had building basements and how engineers sit in their dream-world offices designing but never really building anything. Before long, you will have "saved" thousands of dollars, and he will be building a fancy basement with many walls that will crack and leak. While it is certainly true that some engineers do overdesign (particularly when they are not familiar with underground-house performance), the tendency of the seat-of-the-pants contractor design is serious *under*design. Even if overdesign occurs, the result is a small increase in initial cost, while underdesign means anything from the continuing annoyance of small leaks to serious structural problems. Please remember, if you have any problems at all involving the buried part of the structural shell of an underground house, you will be in for some expensive digging. I have yet to see an engineered design so "beefed up" as to have unnecessary construction costs higher than the cost of just one digging session with a backhoe to fix a minor leak.

This is why I want to give you a little insight into the engineering of poured-concrete structures. While you certainly shouldn't try to do your own engineering without the necessary training, you need to understand what the engineer is trying to do and why. Then you can resist the temptation to just build another basement. If you are going to do your own concrete work, it is even more important to understand what the engineer is trying to achieve, so that his plans are correctly implemented in your work.

What Is Concrete?

Although the words cement and concrete are used interchangeably in general conversation, concrete is the actual building material, while cement is the component in concrete that binds everything together. Concrete has three primary components: cement, aggregate and water. Actually, the main component of concrete is *aggregate,* or the rock part of concrete. In size, aggregate particles range from grains of sand up to multi-inch chunks of rock. In composition, aggregate can be just about any rock, but the characteristics of the final concrete — particularly strength in tension — will be determined to a great extent by the composition of the aggregate. For example, aggregate that absorbs water will result in a concrete that is different in performance from one using nonabsorbing rock.

The aggregate used for concrete should be clean and well graded. That is, there should be a mix of sizes ranging from sand up to a size that should be no larger than one-fifth of the narrowest wall dimension, one-third of the depth of horizontal layers (slabs) or three-quarters of the minimum distance between strands of steel used for tensile strength reinforcement. For our purposes, only natural, non-water-absorbing aggregate should be used. As quarries supplying good aggregate have become depleted, the trend has been to use an increasingly larger percentage of processed aggregate such as pelletized fly ash. Indeed, it is hard to find commercial-batch plants that do not add fly ash as a matter of course, so where high-density, high-strength concrete is needed, as in an underground house, it is necessary to specifically request only natural aggregate in the mix. Lightweight aggregate made from expanded shale, slag, slate, clays or other processed materials should be avoided for the primary structure of the underground house. They offer adequate strength in compression but are weaker than natural-aggregate concrete when the loads place the concrete in tension. Heavyweight concrete (weighing from 200 to over 300 pounds per cubic foot) made with heavy iron ore, barite or even steel scraps can be made, but it is neither economical nor necessary for our purposes.

The cement used in concrete is correctly called *portland cement.* Portland cement is a type of cement developed and patented in England in 1824. It consists chiefly of calcium and aluminum silicates ground into the familiar fine gray powder. Cements other than portland exist, but they are unlikely to be available except on special order. There are also several varieties of portland cement that have different setting times. For underground-house construction, there is rarely a need to go to the extra expense for rapid-setting portland, which reaches design strength in less than half the time of regular portland. Concrete made with ordinary portland cement usually reaches full design strength in about 28 days after pouring. Ordinary portland is available in bulk or in 94-pound bags.

The water used to make concrete should be free of dirt, oils, acids and alkali. Generally, if it is potable (suitable for drinking), it is suitable for concrete work.

It is not uncommon for concrete to have other materials added to the basic three-part mix. These materials, which impart special characteristics, are commonly called *admixtures.* Calcium chloride may be added to cause rapid early hardening to help reduce the chance of damage from freezing temperatures.

Table 5-1 _____

Characteristics of Concrete Admixtures

ADMIXTURE (MATERIAL COMMONLY USED)	BENEFICIAL EFFECTS	ADVERSE EFFECTS
Accelerator (calcium chloride)	Accelerates the strength development of concrete in cold weather and lowers the freezing point of the concrete while setting	Corrosion of steel reinforcing
Air Entrainment (neutralized vinsol resin)	Increases workability and durability of exposed concrete; if used in quantity, a lighter-weight concrete is the result	Reduced strength
Pore Filler (many materials, such as bentonite, kaolin and diatomaceous earth)	Increases cohesive characteristics with no added cement required; used where the added cost of extra cement is not worth the extra strength obtained	Increased drying shrinkage; lower ultimate strength
Pozzolana (fly ash)	Increases resistance of concrete to degradation by sulfates and peaty waters; increases durability of exposed concrete	Reduction in strength when used to excess
Retarder (proprietary chemicals)	Delays the beginning of setting and hardening of concrete in hot weather and during transportation of mixed concrete over long distances	May increase form pressures in deep forms
Superplasticizer (proprietary chemicals)	Considerably increases workability of concrete to allow placement in restricted areas, such as in forms with a high density of reinforcing steel	Requires precise control of mixing process; quite expensive
Water-Reducer (hydroxylated carbolic acid or based on lignosulfonates)	Increases workability in low-water (high-strength) concrete and concrete with aggregates that tend to segregate	Some retardation of setting

Air-entraining agents, which cause the formation of tiny air bubbles in the concrete, are often added to improve workability, increase resistance to damage due to freezing and thawing, and ensure mixing of different-sized aggregate during the pour. Air-entrained concrete is useful for slabs, pavements, sidewalks and other placements exposed to the weather. For the underground house, the poorer structural strength and reduced water resistance caused by the use of air-entraining agents is not acceptable.

Characteristics of Concrete

Concrete is a very complex building material. The mixing ratio is the primary control of strength and water resistance of the cured concrete: how much cement, how much aggregate and, most important, how much water. The actual setting of concrete is a chemical process. The cement changes chemically; it doesn't just dry out. It takes considerable time for this change, called *hydration*, to take place. If concrete is allowed to harden in a moist place, it will continue to harden over a period of years, even though most of the final strength is achieved during the first month. The actual chemical process begins at the surface of each of the tiny grains of cement, where a paste is formed, called the *gel*, which causes the grains to stick to each other and to the aggregate. This gel rapidly stiffens, and within a few hours of the addition of water, the concrete has taken its final set and cannot be significantly changed without physical damage. The time necessary to complete hydration is related to the time necessary for the water to convert each cement grain completely to gel. The rapid-setting portland cement mentioned earlier achieves this because it is composed of unusually fine particles.

The amount of water theoretically necessary for complete hydration is about one-quarter the weight of the cement; however, an additional 10 to 15 percent of water by weight must be present to allow water to reach all the cement particles, so the minimum water-to-cement ratio is 35 to 40 percent. That translates to about 4 to 4.5 gallons of water per sack. Water added over the 25 percent amount will evaporate away and leave minute voids in the hardened concrete. The presence of these voids reduces the strength of the concrete, and the more voids there are, the weaker the final product. As figure 5–1 shows, the ultimate strength of a batch of concrete is dependent on how much water is added above the theoretical minimum. For the underground house, the presence of extra void space in the concrete caused by excess water also reduces the waterproofing characteristics of the concrete. So for high-strength, waterproof concrete, water input should be kept as close to minimum needs as possible. Unfortunately, wet concrete is easier to work, particularly when placing it in deep forms such as walls. It is not uncommon for builders to add water on-site to already-mixed concrete to make it easier to pour. That can cut the ultimate strength of the concrete in half and result in a cracked, leaky structure.

The amount of cement per unit of concrete also has an effect on strength and water resistance. For underground houses, the most common ratios are five or six sacks of cement for each cubic yard of concrete. Figure 5–2 illustrates the relationship between cement quantity and concrete strength.

The size and grading (distribution of sizes) of the aggregate have considerable effect on the strength in tension of the concrete. Better grading generally means higher strength. If all the aggregate is the size of sand, the result is called *mortar*, which has low strength but works easily in thin layers such as between bricks. Strong concrete, which is most likely to be used for underground houses, will have particles ranging from the size of sand up to perhaps 2-inch stone. Generally, the larger gravel is the greatest percentage of the aggregate mass (by weight).

The Mixing Process

The complete mixing of all components into a homogeneous mass is very important for quality results. After the initial mixing, keeping all of the compo-

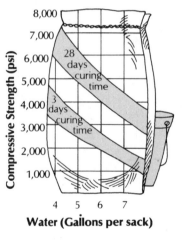

Figure 5–1: The primary determinant of the ultimate strength of concrete is the water to cement ratio.

SOURCE: Redrawn from M. Ferguson, *Reinforced Concrete Fundamentals,* copyright © 1973, with the permission of John Wiley & Sons, New York, N.Y. (Modified from a previously published work by the Portland Cement Association.)

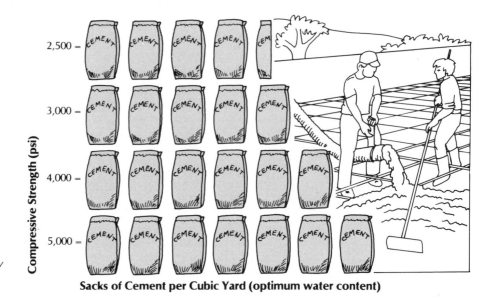

Compressive Strength (psi)

2,500 =

3,000 =

4,000 =

5,000 =

Sacks of Cement per Cubic Yard (optimum water content)

Figure 5-2: This graph shows how the strength of concrete varies with cement content.

nents evenly distributed throughout the concrete mass is a problem. The tendency is for the heavy aggregate to fall to the bottom and a cream of fine cement and water to rise to the top. To keep that from happening, mixing motions must continue until the very moment of pouring. That is the reason for the continually turning barrels on ready-mix trucks. Even the method of pouring can cause segregation of aggregate in the mix, so care must be taken throughout the process (see figure 5-3). Even after placement in the forms, continued agitation of the concrete is good practice. It not only assures a good, even mix but also ensures that concrete will flow to all points and that surface bubbles and honeycombing will be minimal. The agitation is done with manual or motorized tools that are shoved down into the concrete or that vibrate the form walls from the outside. For the type of pours that are done for underground houses, agitation will most likely be performed directly on the concrete. That sounds simple, but don't forget that the bottom of a wall is 8 feet or more from the top, and a forest of reinforcing rods (which have been very carefully placed), electrical conduit, plumbing and who knows what else stands between top and bottom.

If the preparation, mixing and placement of concrete seem difficult, that is simply because they are. Like most things, it is easy to get a result that is fair to adequate, hard to get a good job and very well nigh impossible to achieve perfection. Do-it-yourself pours on this scale rarely make it to good, partly because of the skills needed, which come only from experience, and partly because the nonprofessional rarely has the specialized tools to do a really good job.

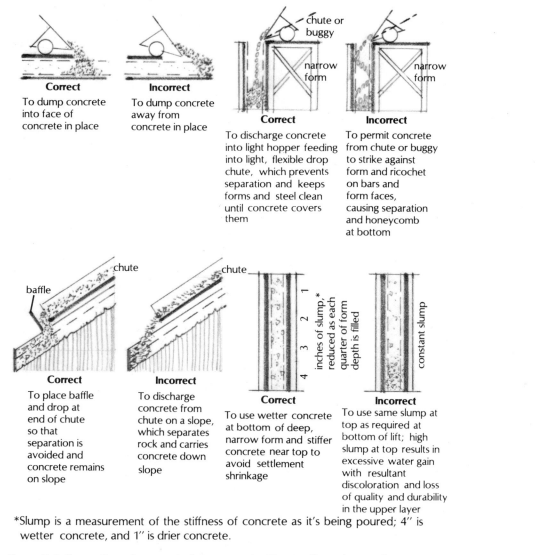

Correct

To dump concrete into face of concrete in place

Incorrect

To dump concrete away from concrete in place

Correct

To discharge concrete into light hopper feeding into light, flexible drop chute, which prevents separation and keeps forms and steel clean until concrete covers them

Incorrect

To permit concrete from chute or buggy to strike against form and ricochet on bars and form faces, causing separation and honeycomb at bottom

Correct

To place baffle and drop at end of chute so that separation is avoided and concrete remains on slope

Incorrect

To discharge concrete from chute on a slope, which separates rock and carries concrete down slope

Correct

To use wetter concrete at bottom of deep, narrow form and stiffer concrete near top to avoid settlement shrinkage

Incorrect

To use same slump at top as required at bottom of lift; high slump at top results in excessive water gain with resultant discoloration and loss of quality and durability in the upper layer

*Slump is a measurement of the stiffness of concrete as it's being poured; 4″ is wetter concrete, and 1″ is drier concrete.

Figure 5-3: Separation of aggregate from cement will occur if care is not taken to ensure that the velocity of all components remains equal during pouring. This illustration shows the correct placement of slab concrete from buggies, the correct placement of concrete on a sloping surface, the correct placement of concrete in the top of a narrow form and the proper way to maintain the consistency of concrete in a deep narrow form.

SOURCE: Redrawn from American Concrete Institute, *ACI Manual of Concrete Practice: Recommended Practice for Measuring, Mixing, Transporting, and Placing Concrete,* order no. ACI 304-73, copyright © 1978, with the permission of American Concrete Institute, Detroit, Mich.

Photo 5-2: Poor mixing, bad forms, segregation of aggregate during the pour, and inadequate tamping and/or vibration of the concrete during and immediately after the pour can result in honeycombing.

Timing the Pour

Ordinary portland-cement concrete mixes should not be kept waiting over 45 minutes. If the concrete cannot be poured within that time, what is known as an initial set occurs and later pouring will result in decreased strength. A very common error is not being ready when the ready-mix truck arrives, which means leaving the concrete on the truck for a half hour or so (after it has traveled a half hour already), and adding water to make up for the stiffened mix. The combination of partial set and added water can only result in reduced strength. Most underground houses are poured in such a fashion as to require more concrete than can be handled by a single truck. The trucks should arrive at intervals designed to permit placement of all the previous truck's load before arrival of the next. It is better to have a concrete crew sitting around for ten minutes than to have the loaded concrete truck doing the sitting. Always assume that some time will be taken in working with steel reinforcing rods (rebar) that have come loose, forms that have moved out of alignment, or other nonscheduled hassles. Space the delivery times far enough apart so that you won't be torn between getting the rebar placed right and getting the concrete in place before initial set occurs.

The Curing Process

For the gel to form, free water must be available to the cement grain. If free water is not available, hydration stops and so does hardening. It is important, therefore, to keep water from evaporating too quickly from the concrete during the curing process. Actually, the strongest concrete is cured under water, believe it or not. Keeping water losses under control is particularly important during the first week of curing, since that is the most rapid period of hardening. To do that, surfaces must be kept moist, out of direct sunlight and shielded from drying winds.

The best days for pouring concrete are calm, foggy days with temperatures in the 60s. A week of that will give you properly cured concrete with few curing cracks. Since you can't count on such "perfect" weather, other strategies are needed. These include using burlap bags that are soaked periodically, moist straw, moist sand, a hose attachment that produces a fine mist, or just about anything you can think of to keep the surface damp. Some patented curing agents are available that are sprayed onto the concrete to retard evaporation. For vertical work in forms, it is easier to keep the surface evaporation down with the forms in place. If the forms are wooden, the inner surface is coated with an oily compound that makes stripping easier and prevents the wood from rapidly absorbing surface moisture from the concrete. If the forms are stripped before a week has passed, it is good practice to keep the exposed surfaces moist.

Because of the high strength needed in the roof concrete, shortcuts should never be taken in assuring proper curing. You can save significantly and ensure a good cure if you personally take on that chore. You don't need years of experience to keep a concrete surface moist, just patience and a supply of water.

Placing concrete in freezing weather is risky. While concrete actually generates its own heat (very thick pours, such as those for dams, actually have pipes carrying cold water through the concrete to keep it from getting too hot), the thin pours in residential construction are not difficult to freeze. Damage due to freezing weather can range from minor surface checking or spalling to severe weakening and collapse. If a pour is made when freezing temperatures are probable, insulating flatwork with straw or fiberglass insulation batts will probably save it. Vertical surfaces are harder to protect, and portable oil- or gas-fired heaters may be necessary. If the shell has lots of openings, they may have to be covered with plastic to let the heat do some good. Frankly, pouring during freezing weather is an "A-number-one" hassle, so avoid doing it if possible.

During the curing process, concrete shrinks, which often causes shrinkage cracks, particularly if curing has been rapid due to high temperatures. These cracks do not penetrate through the concrete and are not abnormal, and if the curing is done properly, there is little likelihood that they will become structural problems. While crack-free concrete is not entirely a mythical substance, it is rare. What takes some getting used to is seeing cracks in concrete — particularly in the roof of an underground house — and not getting worried as long as they do not enlarge.

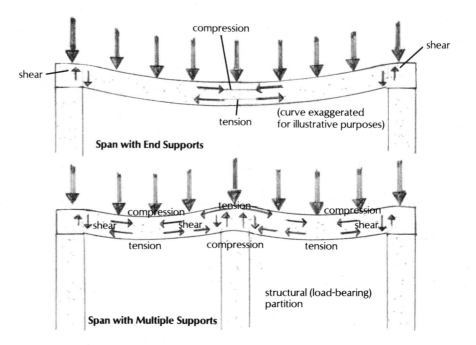

Figure 5-4: Loading a slab places part in tension, part in compression and part in shear stress. The location and type of supports determine the location and type of these stresses.

Structural Characteristics

You've seen it on TV. After a lot of fanfare, the karate expert steps up to the platform, bows, takes a deep breath and with a flash of motion and a loud cry, breaks the concrete slab in half with his hand. Now that's power — or is it? If you have ever worked with concrete blocks, you soon found out just how fragile they are. While they may support tons when placed in a wall, if they are dropped just a short distance onto a hard surface, they crumble. If tapped lightly with a hammer, they break. The great strength of concrete materials under the one situation belies their weakness in the other. What is the difference? The difference is simply in how the force is applied. When placed in compression, the strength of even poor concrete is measured in thousands of pounds per square inch. When in tension, however, wood is stronger than even the best concrete. The reason should be fairly apparent when you think about how concrete is made. Concrete really consists of small rocks glued together by cement. In compression, the rock bears much of the load, while in tension, the bond strength between the rock becomes the limiting factor. If you break a piece of concrete, usually the rock will stay intact, and the break will be along the relatively weak cement bonding boundary. So, while the karate expert is not really faking it, a better test of his skill and courage would be an equally large piece of wood.

The trick in building with concrete is to keep the heavy forces always acting in compression, with only light tension loads. The freestanding vertical wall is just about the only building component that has all its elements in compression. As soon as you put your new concrete roof on top of the wall, subject it to side loads from earth piled up against it, or even allow a good stiff breeze to blow against it, look out. Structures that work primarily through sheer mass, as do some dams, are made of nothing more than concrete, but the overwhelming majority of structural concrete has help in withstanding tension loading by means of cast-in-place steel reinforcing.

Reinforced Concrete

Relatively small steel rods can withstand a great amount of tension, but they are not rigid and will bend due to their own weight if not supported at regular

Table 5-2 _____

Properties of Deformed Reinforcing Bars

	BAR DESIGNATION NO.												
	2	3	4	5	6	7	8	9	10	11	14S	18S
Number of Bars	DIAMETER (IN)												
	0.250	0.375	0.500	0.625	0.750	0.875	1.000	1.128	1.270	1.410	1.693	2.257
	CROSS-SECTIONAL AREA (IN^2)												
1	0.05	0.11	0.20	0.31	0.44	0.60	0.79	1.00	1.27	1.56	2.25	4.00
2	0.10	0.22	0.40	0.62	0.88	1.20	1.58	2.00	2.54	3.12	4.50	8.00
3	0.15	0.33	0.60	0.93	1.32	1.80	2.37	3.00	3.81	4.68	6.75	12.00
4	0.20	0.44	0.80	1.24	1.76	2.40	3.16	4.00	5.08	6.24	9.00	16.00
5	0.25	0.55	1.00	1.55	2.20	3.00	3.95	5.00	6.35	7.80	11.25	20.00
6	0.30	0.66	1.20	1.86	2.64	3.60	4.74	6.00	7.62	9.36	13.50	24.00
7	0.35	0.77	1.40	2.17	3.08	4.20	5.53	7.00	8.89	10.92	15.75	28.00
8	0.40	0.88	1.60	2.48	3.52	4.80	6.32	8.00	10.16	12.43	18.00	32.00
9	0.45	0.99	1.80	2.79	3.96	5.40	7.11	9.00	11.43	14.06	20.25	36.00
10	0.50	1.10	2.00	3.10	4.40	6.00	7.90	10.00	12.70	15.60	22.50	40.00
	PERIMETER (IN)												
1	0.786	1.178	1.571	1.963	2.356	2.749	3.142	3.544	3.990	4.430	5.32	7.09
	WEIGHT PER LINEAR FOOT (LB)												
1	0.167	0.376	0.668	1.043	1.502	2.044	2.670	3.400	4.303	5.313	7.65	13.60

SOURCE: From *Unified Design of Reinforced Concrete Members* by Benjamin Forsyth. Copyright © 1971 by McGraw-Hill, Inc. Used with the permission of McGraw-Hill Book Co., New York, N.Y.
AUTHOR'S NOTE: This table simply lists the physical sizes of the standard sizes of reinforcing bars.

intervals. On the other hand, concrete is very rigid but can't take much tension without fracturing. Put the two together in the right way, and the new, composite material has all the good points of each separate material with virtually none of the bad. The end result has become the most common heavy structural material in use today: reinforced concrete.

While simple in concept, it is not all that easy to produce quality reinforced concrete. First of all, the steel must bond to the concrete; otherwise when the slab is loaded, the steel will just slide and the concrete will fail. To gain that bond, reinforcing steel (called rebar in the trades) is "deformed." Ridges, kinks or other surface irregularities are formed into the steel so that when the concrete sets around the steel it will not slip along the surface. Still, some of the bond is directly between the cement and the steel surface, so the rebar must be clean and free of oil or thick rust. To gain an even stronger tie into the concrete, hooks may be

Table 5–3
Reinforcement-Grades and Strengths

	MINIMUM YIELD POINT OR YIELD STRENGTH, f_y, ksi*	ULTIMATE f_u, ksi†
Billet Steel		
Grade 40	40	70
Grade 60	60	90
Grade 75	75	100
Rail Steel		
Grade 50	50	80
Grade 60	60	90
Deformed Wire		
Reinf.	75	85
Fabric	70	80
Cold Drawn Wire		
Reinf.	70	80
Fabric	65, 56	75, 70

SOURCE: From M. Ferguson, *Reinforced Concrete Fundamentals,* copyright © 1973, reprinted with the permission of John Wiley & Sons, New York, N.Y.
AUTHOR'S NOTES:
*Yield point (f_y) is that tension (per square inch of steel cross section) measured in ksi (thousands of pounds per square inch) above which the steel is permanently stretched. Below the yield point, the steel will spring back to its original length.
†Ultimate strength (f_u) is that tension (also measured in thousands of pounds per square inch of steel cross section) at which the steel will break. For example, two number 6 bars of grade 60 billet steel are used for reinforcing in a beam. Together, their cross-section area is 0.88 square inches. So the maximum tension they can tolerate without permanent deformation is 60,000 × 0.88 = 52,800 pounds. The maximum tension they can withstand is 90,000 × 0.88 = 79,200 pounds.

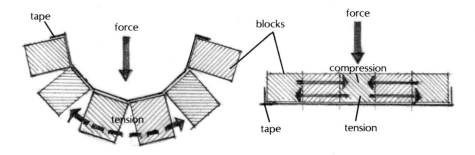

Figure 5–5: Children's blocks can
be used to demonstrate the principle of reinforcing. In the first case,
the tape is in the wrong place, and
the "beam" fails. In the second
case, with the tape (tension reinforcing) on the bottom where the
tension forces occur, the blocks
form a rigid "beam."

formed in the ends of the bar, or sometimes pieces are even welded on so that they stick out like spines and thus keep the bar from slipping.

The second problem in designing a reinforced-concrete slab is determining how strong the steel has to be. When fully loaded, a bar of steel will stretch a little. How much it stretches depends on its physical size and the inherent strength of the steel itself. Although the steel may be a long way from pulling apart, even a modest amount of stretch will result in cracking of the concrete surrounding the steel and, if the stretch is great enough, in a weakening of the all-important concrete-steel bond. Thus, the engineer carefully considers the stresses placed on the steel, and his structural design is aimed at keeping the stretch from exceeding the ability of the concrete to stretch without cracks. For example, beam reinforcement might include eight rebars ½ inch in diameter, each having a tensile strength of 40,000 pounds per square inch of cross-sectional area before permanent stretching occurs.

The third item that has to be carefully considered in design is the proper placement of the reinforcing. The designer has to be constantly aware of where tension stresses occur so that the structural reinforcement is placed in those locations. To get an idea of the importance of the placement of reinforcing, try this little experiment. Place five or six children's building blocks in a row on the floor and tape them together on their tops only. Pick up the two end blocks and what do you get? You get a flimsy, floppy string of blocks. Now turn the whole row over and again pick up the end blocks. Now you have a rigid beam, which, if placed across a pair of additional blocks, will support a very considerable load. As figure 5–5 shows, in the first case, the blocks are placed in tension, and the tape "reinforcing" is not acting structurally at all. In the second case, the blocks are in compression, and the tape is handling the tension load at the bottom of the "beam." That is just what an engineer will do in a concrete structure. He will place reinforcing along the line of greatest tension in the structural member. In a roof slab, for example, the reinforcement is as close to the bottom of the slab as possible while still retaining a solid layer of concrete all around. Usually, it is placed 1 to 2 inches from the slab bottom to ensure that stresses will not break the bond between the rebar and the concrete. Of course, it is not always the slab bottom

*Figure 5-6: Proper placement of
reinforcing requires consideration
of the type and location of
stresses in the concrete. Because
different types of forces occur in
different parts of the same slab,
reinforcement (which counters
tension) must also be placed in different locations.*

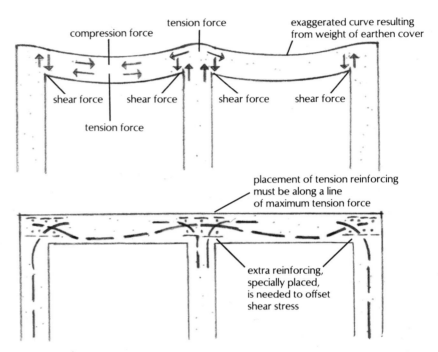

that is in tension; then the bar is placed where tension occurs. Figure 5-6 shows placement in a concrete beam that passes over a column. The top of the beam is in tension over the column, while the bottom is in tension in the span, so the reinforcing is placed accordingly.

The type of reinforcement so far discussed is called *tension reinforcement.* A second type of reinforcement necessary for structural slabs is *temperature reinforcement.* Temperature reinforcement is placed at right angles to tension reinforcement and provides resistance to temperature and shrinkage stresses. While it has considerably less to do with structural strength than tension reinforcement, temperature reinforcement has a lot to do with limiting cracking in slabs, and therefore it has a great effect on how waterproof a slab remains. Temperature steel is usually placed on top of the tension steel, since it's important to install it at the middle of the concrete slab because the stresses it compensates for are generated by the mass of the concrete itself rather than by the loading placed on the structure. Also, temperature-reinforcing steel is usually much smaller than tension-reinforcing steel. Welded-steel wire mesh is often used for temperature reinforcement. Note that the mesh designed for concrete reinforcement is more carefully made and of a considerably better grade of steel than welded-wire fencing, which may look very similar. The tensile strength of steel for reinforcing is high and carefully controlled because unless that strength is a known constant throughout a batch of steel, reliable structural design is impossible. While you may know of a house that was totally reinforced with used

Photo 5-3 (above): *With cast-in-place concrete roof systems, conduit, reinforcing steel and temperature steel must be shaped and placed carefully for maximum strength.*

Photo 5-4 (right): *Care in the placement of steel, electrical conduit and forms is required for quality results. Note the bends of the wall steel that tie the wall and roof together.*

barbed wire, don't reinforce yours that way. If you do, I will still come to visit you, but let's stay outside on the patio.

Special Structural Problems

Besides compression and tension stresses, which occur along the length of concrete slabs, there are also high concentrations of stress at a few other sites across the slab. One type of site generates what are called *shear stresses*. A region of particularly high shear stress occurs in a slab roof right over a freestanding column. The amount of weight supported by that column is many tons, and the tendency is for the column to "punch" through the slab rather than to hold it up. Special reinforcing at this very high stress area is needed to transform shear to compression and tension stresses, which can be handled more easily. Likewise, where roof slabs are supported by walls, high shear forces are present. Again, special treatment of reinforcing at such sites saves the day. One particularly touchy place is where a floor joins a wall. In some parts of the country where soils can accept high loadings and are dimensionally stable over a wide range of conditions, it is not unusual for a floor slab to be poured continuously with footings and with the vertical walls poured over that. While that may be all right in some aboveground frame houses, it is inviting trouble in the underground

*Photo 5–5: Because of the large
number of steel reinforcing bars
and the importance of their position for cast-in-place roof systems,
care in placing the concrete is of
great importance.*

house because of the great weight of the structure that must be supported by
the footings. The shear stresses on the perimeter of the floor slab become
enormous as the soil under the footing yields over time. The result can be serious
cracking around the floor perimeter with resulting loss of waterproofing and
even the tilting of parts of the floor. To keep that from happening, it is best to
have a floor slab that is not rigidly attached to the wall. The obvious problem with
such an arrangement is the maintenance of perimeter waterproofing. Several
waterstop systems are available to fill that need, and these are discussed in
chapter 6.

Wherever sharp inside corners occur, stress concentrations are present, and
additional reinforcing is needed to prevent cracking. Extra reinforcing is called for
around roof openings, window and door openings, floor drains and all intersec-
tions between slabs. Stress concentrations can be reduced by filling in corners
with extra concrete called a *fillet* or *cove*. This is particularly useful at the base of
parapets, retaining walls and other slabs that have a turning action exerted against
their base. Also, making openings round instead of with sharp inside corners is
helpful. A round skylight well or chimney penetration will be less likely to crack
than a square one, although both need additional steel around the opening.

Your engineer should create a design that will keep all these potential
problems from becoming real ones. All you and your builder have to do is follow
the engineer's plans as to concrete mix, reinforcement strength, size, and
placement, and concrete thickness. If you do that properly and allow the
concrete to cure in a proper manner, you should have a reinforced concrete
building that will retain its structural and watertight integrity for many decades.

Strength with Economy

In a structure made up of loaded slabs, like the walls and roof of an underground house, the distance between rigid supports, or the clear span, becomes the main factor determining the thickness of the concrete and the reinforcing required to resist the stresses. The closer the supports, the thinner the slab and the less reinforcing is needed. Naturally, the cost goes down fairly rapidly as the strength requirements are reduced. For that reason, internal walls of underground houses are often masonry in order to provide structural support for the roof and to act as a buttress against wall loadings from the earth. In real designs, some large open areas are needed. For example, in figure 5–8, one end of the house has a number of small rooms, allowing for short roof spans if partitions are masonry and load bearing. The other end of the house, however, is basically a single open space. If a column can be placed near the center of that space, considerable savings can be realized in the roof structure. If you insist on a completely clear-span roof over the large room, you can consider having a thicker, more heavily reinforced slab over the one large room. By using the walls in the other end of the house as structural supports, you can use a thinner, less costly roof slab over the rest of the house.

In any case, you should ask your engineer to advise you as to the trade-offs between structural interior walls and an essentially clear-span roof. Under some soil conditions, differential settling of noninterconnecting footings will occur. Such sites should not have lots of interior columns and short interior supporting walls, because they will settle at a different rate from other parts of the house and will probably contribute to cracking instead of preventing it. For other sites, making all interior walls structural will reduce the cost of the exterior shell to the extent that the overall cost of the house will be lower than the cost of one with clear-span walls and roof. A qualified engineer will consider these various structure/cost trade-offs and will work with you to create the most house for the least money. I guess it is appropriate to voice again the warning that any given structural design is specific for a particular site and set of conditions and should never be considered for construction anyplace else without an engineering review.

Placement of Reinforcing Steel

Since the final location of reinforcing steel is very important to the structural performance of the concrete, steel must be well secured in the concrete forms before the pour. Considerable force results from liquid concrete flowing down a vertical form or rushing horizontally on a roof or floor. If the reinforcing steel is not well anchored, it will move from its optimum placement point to some other not-so-optimum location. You then have a weakened structure. There are many books that go into the details of form building and steel placement, and they or an experienced concrete tradesman should be your reference. I will risk being called a fussbudget, or worse, by once again reminding you that if everything is not right at the time of the pour, it will be hard to change in just an hour or so. To

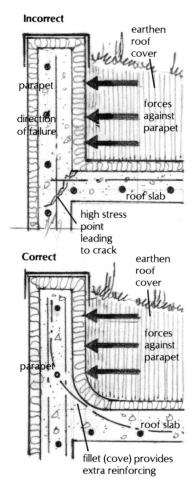

Figure 5–7: Sharp corners, such as where a parapet joins a roof, are areas of high stress, and extra reinforcing should be added. The addition of a fillet, or cove, of concrete at the inside of such a corner distributes stress over a greater surface and allows better placement of tension reinforcing.

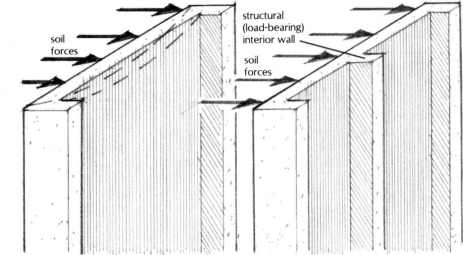

Figure 5–8: In underground structures with roof slabs that don't strengthen walls, internal buttressing with structural partitions may be cost-effective. The room on the left is a single open space with no interior buttresses, so thicker walls and heavier reinforcing are required to resist soil forces. The room on the right has an interior structural partition that allows for lighter construction.

coin an understatement: *Concrete sets hard.* Anything that is forgotten or done incorrectly relating to the concrete pour will become a lasting memorial to someone's inattention or lack of skill. My friends tell me that I am not entirely rational on this subject. My excuse is that wielding a 20-pound sledge or a pneumatic hammer to break up all those concrete pours I have had go wrong may have scrambled a bit of the old gray matter.

Cast-in-Place Concrete

Probably the most common use of concrete is to cast or pour it into forms. That is, after the concrete sets, it is not moved. Obviously, that means that the forms have to be placed exactly on the site so that when they are removed, the house shell is complete. Within the generic system called *cast-in-place* there are also some subsystems. The *monolithic-pour* system places all the structural concrete in one single pour. The footings may be poured in advance, but the walls and roof and any other structural concrete components are placed at one time, so there are no cold joints. A *cold joint,* sometimes called a *construction joint,* occurs anywhere concrete is poured in contact with previously cured concrete. Cold joints can be a leakage site and a possible zone of structural weakness, since fresh concrete bonds poorly to cured concrete. For those reasons, the monolithic pour has the potential of being the strongest and most leak resistant of all the cast-in-place systems. I say "potential" because a lot can go wrong, too. Placing all that concrete in a continuous pour requires the careful scheduling of a space shot, workers with the endurance of marathon runners, and a lot of confidence. Once the sequence is started, there is no graceful way of stopping until the last yard is poured. The result may well be worth the effort, though. A well-designed, properly poured monolithic shell is a great start for an outstanding underground house. But, believe me, it is not for the faint of heart!

For most of us, the much easier method is the *sectional-pour* process. The shell is divided into major structural sections, each of which is poured independently of the others but joined together by extensive reinforcing steel ties. The cold joints that result are made waterproof by installing waterstops. *Waterstops* are flexible strips placed so that concrete on both sides of a cold joint is poured around part of the stop (see figure 5–10). The flexibility of the material, the slight shrinkage of the concrete as it cures and the long, usually wavy, path that water would have to take to get across the waterstop barrier all combine to make the cold joint as waterproof as if it were a continuous pour. Although some brave (dare I say foolhardy?) souls don't use waterstops in all cold joints in underground houses, I can give you the names of several owners whose carefully considered opinion is that such builders should be lightly tarred and gently feathered prior to being dropped into the water seeping into the houses they built. When using the sectional-pour method, have your engineer recommend the best stopping places for the pour. Cold joints are weaker than continuous pours and should be placed where the structural design will not be compromised.

Sectional monolithic pours may be practical in some houses, particularly houses that have one very long axis and the other short. The house is divided into sections along the long axis, usually about 30 feet each, and each of the sections is poured as a monolithic unit, with sections tied together by steel reinforcing. This allows the designer to use the structural advantages of a monolithic pour while keeping things on a scale that is more easily manageable. Having a cold joint every 30 feet or so is not a bad idea anyway. Aboveground concrete buildings have these construction joints at regular intervals to allow any expansion and settling stresses to be relieved. In the underground house, temperature changes are minimal and stresses tend to be pretty constant, so these joints are not as necessary, but they will still help keep walls from cracking. Naturally, waterproofing of these joints is necessary, and waterstops plus flexible waterproofing compound should be installed.

Concrete Forms

Concrete forms should be rigid and have a cross attachment between adjacent panels. In basement work, rough joints and bulges in the finished concrete are not too worrisome. You, however, are not going to be pleased if the ceiling of your house looks wavy or the walls turn out with seams that have to be individually chiseled and ground flush. Ceilings, in particular, are tough problems for form builders. While it is possible to site-build wooden forms that will provide flat, fine-seamed roof slabs, the expense is great. Much to be preferred are the prefabricated shoring systems made for this purpose. They are often available through construction-equipment rental companies in the larger cities. But just because they're rented, don't get too anxious to take them down right after the pour. The extra few days' rental is a lot cheaper than replacing a

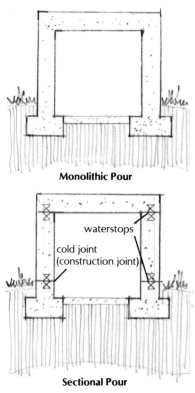

Monolithic Pour

waterstops

cold joint
(construction joint)

Sectional Pour

Figure 5–9: In a monolithic pour, all of the structural concrete is poured at one time. Sectional construction is simpler, but cold joints (construction joints) require special waterproofing attention. A cold joint occurs where concrete is poured in contact with concrete that was previously cured.

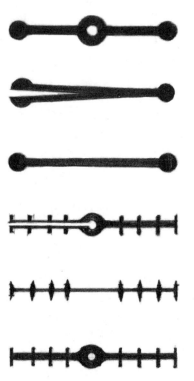

Figure 5–10: Waterstops (shown here in cross section) are manufactured in a variety of sizes and shapes, but all are used to prevent water entry through cold joints.

SOURCE: Redrawn from James J. Brown and Terry A. Johnson, *Earth Shelter Waterproofing: The Guide to Total Moisture Control*, copyright © 1980, with the permission of Underground Homes, Portsmouth, Ohio.

roof made structurally unsound by having to support its weight before it has cured to a sufficient strength. As in all technical decisions about strength, follow your engineer's advice regarding when to remove the forms. Err on the long rather than the short side, if err you must.

Form design and construction is a trade in itself. Before tackling this specialty yourself, be sure to learn as much as possible from form builders and books so that you will know what you are really getting into. You simply cannot get good concrete unless you have good forms that stay in place, that don't bulge and that can be easily stripped after the concrete cures. Don't let your first form be like mine. I was forming a mere 4-foot retaining wall for a truck-loading dock. In making the form, I used what I considered excessively strong 2 × 6 construction with a veritable forest of braces going every which way. Even so, the braces broke, the form collapsed, and my education was advanced one more increment.

Not all forms are removed, however. It is not uncommon in commercial work to use a corrugated metal form that is left in place and becomes the exposed wall or ceiling surface in the completed building. I have seen only one house where this was done, and the result was pleasing because an attractive corrugation pattern was chosen. Finishing the metal is usually much easier than finishing bare concrete. Of course, the metal acts as a water barrier, too.

Prestressed Concrete

While cast-in-place reinforced-concrete construction is the most common method for constructing an underground house shell, other approaches are possible. One that is gaining favor with builders is *prestressed concrete*. The principle behind prestressing is simple: Concrete has little strength in tension but a lot in compression. In prestressing, the concrete is placed under such a strong compressive stress, through the use of stretched, high-tensile-strength steel tendons cast within the concrete panel, that the design loads for the concrete are never great enough to place the concrete in tension. Remember that regular reinforced concrete allows tension stresses to be applied on the concrete, so even though the steel reinforcing ultimately takes up the stress, enough stretch is allowed that very small cracks occur. Fully prestressed concrete should not crack all the way through a panel, and if it does, it may be the signal that it is near failure. Sometimes the steel is stretched only partway to its maximum, and considerable strength is available for additional stress. Then the concrete is *partially prestressed,* and its cracking characteristics lie between regular reinforced concrete and fully prestressed material. Because cracks are a common path for water entry into an underground house, fully prestressed concrete is usually used to keep cracks to an absolute minimum.

Let's go back to the taped-block experiment, but this time instead of putting a strip of tape along one side, wrap the column with a rubber band. Again, place

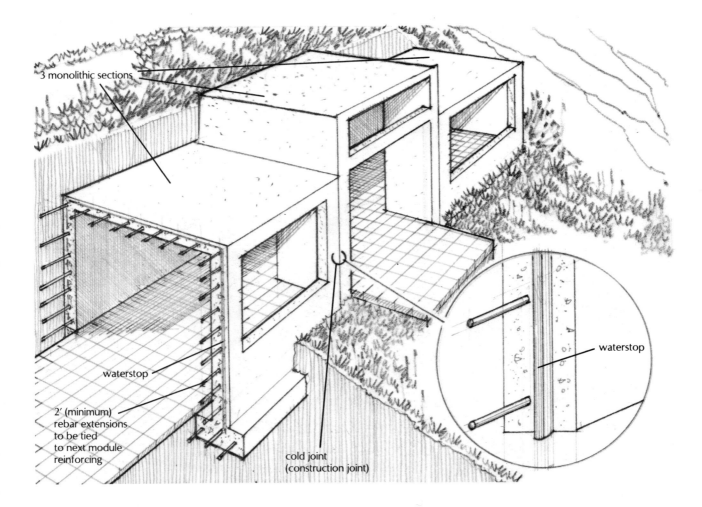

3 monolithic sections

waterstop

2' (minimum)
rebar extensions
to be tied
to next module
reinforcing

cold joint
(construction joint)

waterstop

Figure 5-11: Monolithic sections may be joined together with rebar. Waterstops provide resistance to water at module construction joints.

the "beam" you have created on a couple of blocks and note how strong it is. Now add several more strong rubber bands to the one already there, and notice how much more pressure the "beam" will take before it "cracks." That is prestressing. The stronger the bands, the more prestressing is present and the more weight the beam will support before the blocks begin to separate.

Two common approaches to prestressing concrete are *pretensioning* and *posttensioning*. The *pre* and *post* refer to whether the steel is stretched before or after the concrete sets. Pretensioned panels are made by stretching the high-tensile-strength steel tendons between fixed anchors and then pouring the concrete around them. After the concrete hardens, the steel is released, and its tension is transferred to the concrete along the entire length of the tendons. In posttensioning, the tendon does not bond to the concrete and is not stretched

Photo 5-6: Easily installed modular steel roof-form supports may often be rented for much less than site-built form structures.

Figure 5-12: For short spans, corrugated steel may be used to form the underside of a roof slab and then left in place as the ceiling surface after the concrete cures.

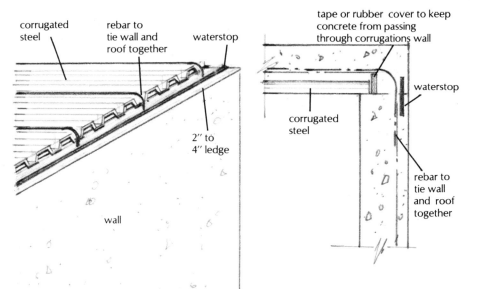

corrugated steel

rebar to tie wall and roof together

waterstop

2" to 4" ledge

wall

tape or rubber cover to keep concrete from passing through corrugations wall

corrugated steel

waterstop

rebar to tie wall and roof together

until after the concrete hardens. After stretching, anchors are put in place, which clamp the tendon to the concrete, and when the tension is released, the anchors transmit the stress to the concrete.

Pretensioned, Prestressed Concrete

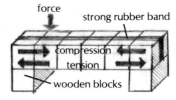

If you stay in budget motels when you travel, you have probably noticed that the ceilings are often made of wide concrete "planks" rather than one smooth surface. Those planks are almost certainly *pretensioned, prestressed, precast* concrete panels. They were made in a factory, delivered to the building site and laid across the end wall supports just like boards on a scaffold. If you could see the ends of the panels, you might be surprised to see that they are hollowed out with a series of large holes running the length of the panels. These panels are made by stretching a series of high-tensile-strength steel wires, cables or rods through a concrete form and pouring high-strength concrete around them. When the concrete sets and bonds to the steel, the stress is released on the tendons, and that compressive load is transferred to the concrete. This compressive force causes the panel to buckle slightly and increases its strength to equal the strength of a regularly reinforced concrete panel that is nearly twice as thick and almost three times as heavy. The precast, prestressed panel is also relatively light because of the holes along its length. While full panel thickness is needed for bonding to the tendons and for reducing flexure, perforating the panel between the tendons does not substantially reduce its ability to withstand loads. These same holes are also handy as plumbing, wiring and ventilation channels.

Precast, prestressed panels have long been used for commercial construction but rarely for houses, where such strength is usually not needed. But strength *is* all-important to underground construction, where the roof commonly has to support over 300 pounds per square foot.

Several underground builders have realized the potential of precast, prestressed panels, at least where local panel manufacturers can deliver and install their roof units for less than the cost of cast-in-place concrete. Why less? As it turns out, there are several cost-saving advantages to working with precast, prestressed panels for roofs. The major costs for cast-in-place roof systems — on-site form construction, shoring and steel placement — are borne by the panel manufacturer, who spreads that cost over many panels. While cast-in-place concrete may require weeks of curing before the roof can be backfilled, the panels arrive at the job site fully cured and ready to take full loading, saving a lot of construction time. In addition, cast-in-place construction requires detailed engineering, skilled labor and expensive tooling, while panel installation can be done with a few strong backs, a rented crane and a foreman. Finally, the quality control of the concrete mix and the pour of the precast, prestressed panels is usually much better than what can be achieved in the field. That particularly suits areas where it is difficult to get good concrete work done.

Figure 5-13: Children's blocks can be used again to illustrate prestressing. When the prestressing (the tension of a rubber band wrapped around the blocks) is greater than the tension force, the "beam" is rigid.

Photo 5-7: Precast, prestressed concrete panels permit quality control and rapid construction of houses.

Photo 5-8: Precast concrete panels form the walls of this underground house under construction.

One obvious problem with precast, prestressed panels is the presence of joints between the panels, commonly every 2 to 4 feet. Although the edges of the panels may have an interlocking tongue-and-groove design to keep them aligned, the crack is still a potential leak site. Commercial construction practice is to pour a 2- to 4-inch layer of concrete over the panels. This slab is also reinforced with welded-wire mesh. That both increases the strength of the roof at the panel joints and provides a much more waterproof surface. Some builders of underground houses follow that practice, but many do not. Those who leave off the covering layer of concrete do use extra care in applying the water-proofing to the roof. The joints are filled with waterproofing mastic, and the surface is treated with an overall waterproofing system.

Even with all their advantages, precast, prestressed panels still are in the minority when it comes to underground-house roofs. There are several reasons for that. First and probably most important is the fact that there are many home-scale builders with the equipment and skills for cast-in-place construction but very few with experience in using panels. Second, you have to be within reasonable delivery range of a manufacturer and have access to a crane to obtain and install the panels. Expensive manufacturing jigs and forms are needed to make precast, prestressed panels, and manufacturers are rarely found outside of major metropolitan areas. There are exceptions, however. In some areas, local concrete and labor is so expensive that delivery of precast units in excess of 200 miles is still profitable. The only way to be sure is to carefully price both cast-in-place and panel roofs.

The design constraints that are caused by the use of standard-sized, prefabricated panels are an additional reason for preferring cast-in-place over precast, prestressed panels for some houses. Special sizes or nonrectangular

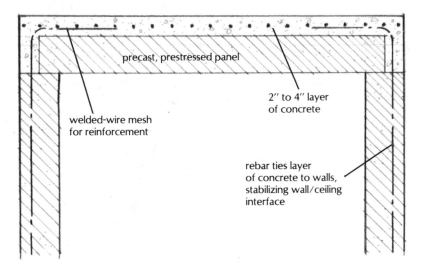

precast, prestressed panel

welded-wire mesh
for reinforcement

2″ to 4″ layer
of concrete

rebar ties layer
of concrete to walls,
stabilizing wall/ceiling
interface

Figure 5-14: Casting a thin concrete slab over precast, prestressed panels provides a solid surface for waterproofing, increased structural strength and a superior tie between roof and walls.

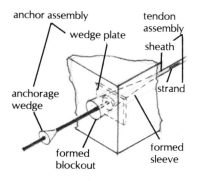

Figure 5–15: Anchor systems for posttensioning cables depend on wedges and wedge plates placed in the concrete. The wedge shape provides great holding power, which actually increases as the cable stress is increased.

SOURCE: Redrawn from Portland Cement Association, *Constructing Earth Sheltered Housing with Concrete,* 1981, Portland Cement Association, Skokie, Ill.

shapes require special engineering and nonstandard forms. Large holes, such as those for skylights, are next to impossible to cut and should be formed into the panel at the factory (although small holes, such as those for plumbing vents, can be cut on-site). Precast, prestressed panels are a most advantageous choice when the site is within about 50 miles of the plant and when used in a predominately rectangular design with no extraordinary roof penetrations.

Precast, prestressed panels can also be made for walls, but they require a special design. Roof panels have a slight camber (curvature), which is lost when the slab is put under load. In a wall, the camber remains. A reinforcing design has been developed that results in flat panels, but few manufacturers are making them, since the demand is still small. For walls, furthermore, there are few advantages in using panels. Skilled concrete workers who can handle wall-form systems are relatively inexpensive to hire. Special engineering design and fasteners are required for precast, prestressed wall panels in order to join walls, roof and floor together in an adequately strong structure. Without this special attention, particularly at corner joints, you could end up with a "house of cards." One approach to the design of wall systems is to make the panels as long as can be transported and to lay them on their sides rather than the more conventional vertical placement common in commercial construction. Even under the best of conditions, however, walls for underground homes made from precast, prestressed panels are still very much experimental and generally cannot be considered cost competitive with cast-in-place walls.

Posttensioned, Prestressed Concrete

Since pretensioned, prestressed concrete construction is clearly not practical for on-site fabrication, field prestressing must be done with posttensioning techniques. Until recently, posttensioning was in the domain of the commercial/industrial builder. Structures ranging from mammoth water tanks to huge free-form concrete buildings use posttensioning. One builder, Lon Simmons, in High Ridge, Missouri, began using the system for his underground houses and quickly became sold on posttensioned concrete. He enthusiastically notes the reduction in material used, the relative freedom from cracking and the ease of waterproofing that result from using posttensioning.

Several different methods are used for installing the tendons. The first, developed by the famous French engineer Eugene Freyssinet, uses high-tensile-strength wire or cables inserted into ducts that have been formed in the concrete. The cables are stretched with hydraulic jacks to about 80 percent of their capacity, and wedge-action anchors are clamped on the tendons at the concrete surface. When the jacks are released, the anchors grasp the cables firmly and push against the edge of the concrete with tens of thousands of pounds of pressure. Grout is then forced into the space left around the tendon to prevent corrosion and to bond the steel to the concrete over the length of the cable.

Simmons uses a modification of that original method, in which the tendons are covered with a plastic sleeve. When this plastic-coated steel is cast into the concrete, there is no bond formed between the steel and the concrete, but neither is there any gap between the steel and the concrete that has to be filled with grout. Thus, after the slab hardens, the steel is stretched and anchored, and only the anchors and the end of the cable must be protected from corrosion.

There are still other proven systems for posttensioning. One uses a sulfur-coated wire cast into the concrete. The wire is then electrically heated, which both melts the sulfur and causes the wire to elongate. While still hot, the wire is anchored to the edge of the concrete. When the wire cools, it shrinks and becomes stressed, and the sulfur solidifies and bonds the wire to the concrete. In the USSR, tendons are often placed *outside* the concrete in cast grooves, then stressed and grouted — a system remarkably similar to our rubber-band-and-block model.

A nice aspect of posttensioning is that the tendons are easily cast into curved positions, allowing prestressing of domes and corner panels and stressing with a built-in uplifting force for ceiling panels. Lon Simmons runs tendons in a grid pattern through the walls, the floor and the roof so that when stressing is complete, a very rigid and essentially crack-free concrete box results, even though the weight of concrete and steel is less than that used for a regular reinforced concrete house. Add to that the fact that clear spans of 28 feet and more are practical due to the high strength of the prestressed roof, and it is clear why Simmons thinks posttensioned concrete is the answer to an underground builder's prayers.

There are still few builders using posttensioning, however, because it is still a new technique for residential construction. The few contractors with postten-

Photo 5–9: This equipment is used in residential posttensioning systems.

Photo 5–10: After the concrete sets, the cables are tensioned with specially designed hydraulic jacks; wedge anchors are installed, and the jack is released, leaving the cables in high tension.

sioning experience are usually specialists in commercial/industrial construction, and they are not involved in residential systems. Also, better quality control is necessary over the concrete itself, which is not always possible with local batch producers. The concrete itself needs to be of a high strength due to the high local stresses placed on the material by the tensioning tendons, so a rich mix with minimum water is necessary. That in turn means that more vibrating and tamping are needed to properly fill all the voids in the form, so a very stable form system is needed. Considerable care has to go into the placement of the steel tensioning members, or the effect sought by the engineer will not be attained. Oh yes, *each* house design *must* be engineered, and while there are many engineering firms that have worked with large-scale posttensioned concrete structures, few are interested in home-size design. Simmons was able to find local, experienced engineering service, and you may also, but some searching may be required.

Prestressed Systems and the Owner-Builder

With precast, prestressed panels for roof members, owner placement can be done as long as you are willing to harass the manufacturer into giving you a short course in the installation of his panels. You must provide the manufacturer with a good estimate of the loading over the panels (primarily backfilled soil) and how it is to be distributed along their length. You must be *sure* that you and the

Figure 5-16: If the precast, prestressed panels are a few inches short and thus create a ledge at the roof/wall corner, a concrete layer can be poured flush with the vertical wall surface. To prevent concrete or water entry into the hollow cores of the panels, the ends can be covered with wadded paper or roll roofing.

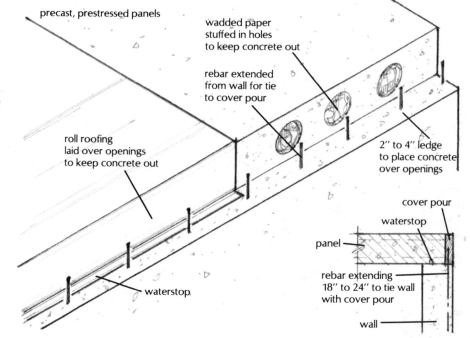

manufacturer are speaking the same language when you are discussing panel length. Clear span to you may mean the interior separation of supporting walls; to the manufacturer it may simply be the total length.

Also, for motels and similar buildings, the panel length is usually measured from the outside of one supporting wall to the outside of the other. For your underground house, you will probably want to order the panels 2 or 3 inches shorter. That way you can stuff some newspapers tightly in the holes at each end and spread concrete over the ends flush with the vertical wall surface. If you leave the ends open, you probably will have trouble keeping water out of the holes, and someday might find your ceiling damp in stripes along the length of the panels! If you do buy panels that come flush to the outside of the walls, a metal flashing to cover the holes will keep your insulation and waterproofing from being forced into them when the site is backfilled.

Most of the problems encountered in using precast, prestressed panels are the result of poor planning and project management. The walls must be absolutely ready for panel installation and attachment. Be sure to find out exactly how the panels will be attached to the walls before the walls are built! Some attachment systems call for metal plates that must be accurately cast into the wall tops for bolting or welding. Others use no fasteners at all but depend solely on gravity to "keep the lid on." This gravity "attachment" is an approach not recommended for underground roofs, since it is likely that backfilling operations will cause movement of the unattached panels. Whatever the attachment system, the wall tops must be reasonably flat and quite level so that the panels will be supported over their entire width. To this end, many manufacturers recommend that the top of the wall be "buttered" with fresh mortar just prior to panel installation. If that is the case, a specific maximum of mortar thickness will be recommended. Some manufacturers call for a neoprene bearing pad to be placed between the wall top and the panel bottom. Note that this is *not* intended as waterproofing but serves to distribute the roof load over as much area as possible. It is important that the mortar bed be of uniform thickness along the wall. If the panels are not all at the same level, appearance and joint strength will suffer and waterproofing will be more difficult.

Summary of Precast Panel Construction

Get the system engineered by the panel manufacturer or an independent firm or both. Be sure that you and the manufacturer are talking about the same thing when dimensions are discussed. Find out everything about the roof panels and how they are to fit to the walls before you start the walls. Make sure that any attachment devices are properly placed when the walls are poured.

(continued on next page)

Summary of Precast Panel Construction—*Continued*

Have *everything* ready on the day the panels are to arrive: mortar for the bedding coat if needed, a properly sized crane for lifting the panels off the truck and into final position (make sure both you and the crane operator know the manufacturer's recommended lifting procedure *for the panels you ordered*) and enough manpower to work both ends of each panel for final positioning. Be sure the panel camber is placed in the direction the manufacturer recommends (almost always with the camber up). Be ready to complete all necessary attachments before the mortar bed hardens.

Again, the most common problem is not having everything ready when the panels arrive! It will be costly for you to have everyone standing around, plus the result may be a poor installation that will provide you at best with a building you are not quite happy with, and at worst with one that is unsafe. Also, you don't want to have a $150-per-hour crane sitting around while someone goes to find some mortar for the bedding.

Posttensioning and the Owner-Builder

Posttensioning is usually a poor choice for the owner-builder. Proper placement of steel is *very* important, and the engineering drawings are usually not very easy to read. Special commercial-quality concrete-placement equipment is needed to work the pour. Hand "jitterbugs," shovels and similar hand tools just will not properly place low-water, high-strength concrete. The jacks and fixtures necessary to accomplish the actual tensioning are not generally available for rental and are too costly to purchase for one job. Training in tensioning procedures is needed but hard to get. Perhaps you *can* place the steel correctly; perhaps you *can* build forms of the necessary rigidity and accuracy; maybe you *can* properly place the concrete; but you will still have to get someone to come and pull the steel. Chances are, if you do everything else, you will find it extremely difficult to find a contractor willing to do just the stressing. My recommendation is to get someone with experience to do it all, and on that basis, here is a summary of planning steps for posttensioned construction.

Summary of Posttensioned Construction

Have the building engineered by a firm that has experience with small-scale designs. You can expect most structural engineering firms to be able to do the work for you, but if this is their first concrete house or their first posttensioned job (and both are likely to be the case), you may have the pleasure of paying for the time necessary for the engineer to figure out how to do the design as well as doing the design itself. If you can, find an engineer who has already had this training paid for.

Find a competent, trustworthy contractor who has experience in home-scale posttensioning. Make sure you and your contractor have an ironclad agreement that your engineer is the *final* authority for all questions.

Have your engineer come to the site and check all form work and steel placement. Your contractor may not be very fussy about tying the steel in place. If he isn't, the stiff, dry concrete that he should be using may cause significant shifts of the steel. If the job looks wrong or shabby *get the engineer out there.* Do not be afraid to squawk loudly. You have to live in that thing!

Visit the concrete supplier and make sure he knows what mix is required. In your written agreement (contract) with the contractor, be sure that there is a clause that allows you to take samples at the beginning, middle and end of the pour from each truck and that also indicates that if they do not live up to specifications, he is responsible for doing the job over. Why so much fuss about the concrete? In construction, one of the least understood and most often "fudged" areas is concrete. With prestressed systems, you just can't fudge and get away with it.

If possible, your engineer should be at the site for the pour. If not, look at the consistency of the concrete. If, as it comes off the truck, it piles up and flows poorly, it is probably a good mixture. If it quickly runs to the far corners of the forms and sits with a glistening watery surface, you are in trouble.

Do not accept concrete that has calcium chloride added. Most manufacturers of posttensioning steel get upset if their steel is stressed in concrete with a corrosive calcium chloride additive. Contractors often request calcium chloride added to the mix for faster setting in cold weather, but unless you get a clear statement to the contrary from your steel dealer, do not use it.

Obviously, if you can find an experienced contractor you can totally trust, you can leave the site and come back when it's all over. Despite all the fuss I have made, there really are competent people out there. The key is to look for experience and good recommendations from past customers.

Ferrocement Construction

Making concrete structures that have surfaces that curve in more than one direction — domes, for example — is very difficult to do with conventional forming methods, but construction methods using *ferrocement* can create almost any shape. Ferrocement is a type of reinforced concrete and gets its name from the Italian word *ferrocemento,* which was the term given to this type of concrete when it was developed in the 1940s. The term ferroconcrete refers to reinforced concrete in general. In order to produce a ferrocement structural shell, concrete is forced into the layers of a wire-mesh *armature* that usually includes strategically placed steel reinforcing bars (rebars) and possibly a welded-wire fabric. The strength of the resulting shell depends on the thickness of the concrete and its composition, the placement and strength of the rebars, and the effect of the curvature in offsetting outside stresses.

The construction of boats using ferrocement technology probably popularized this construction method. Although ferrocement boatbuilding dates from at least 1872, it has only been since World War II that significant numbers of boats have been built of concrete. Present-day technology allows hull thicknesses of less than 1 inch, so some of those beautifully sleek boats you see in the harbor may be made from concrete! Numerous amateur boatbuilders have also created their own concrete hulls. The complex curves of a boat hull can be very expensive to create if only one hull is to be made with the shop forms, bending rigs and other paraphernalia necessary when working with conventional boat-hull materials. The wire-mesh armature for a ferrocement hull is relatively easy to shape into the most complex hull contours, and it requires little in the way of special rigging. As a result, the construction costs of a single hull of ferrocement can be significantly less than even a commercially made hull of other materials. Lower costs may only be possible to the backyard boatbuilder, however, because ferrocement construction is very labor intensive.

The advantages of using ferrocement to make boats also apply to house construction. The material allows for the creation of complex curves with

Photo 5-11: The armature of a ferrocement structure is a combination of heavy- and fine-gauge steel reinforcing. Shown here is the Pearcey house under construction (see chapter 8).

relatively low material costs, although labor costs can be high. There are several free-form aboveground houses that incorporate the thin-shell, ferrocement-boat technology, but boatbuilding techniques are generally inadequate for building underground because of the need for greater rigidity and higher strength. Boat hulls are designed to flex in response to rapidly changing forces of waves and wind, but house shells must be rigid to withstand the constant, massive pressures of the earth. As a result, underground houses made of ferrocement may require greater thicknesses, heavier reinforcing and multi-layer construction. The structural design of complex curves may call for engineering fees that may be considerably more than those for houses with flat surfaces. The exception is for houses with domes or arches, since these simple curves are reasonably easy to analyze structurally. Considering the increased thickness needed, the heavier reinforcing and the higher engineering costs, the cost of ferrocement approaches may exceed that of conventional concrete construction except when truly free-form shapes are desired. If a design calls for flat walls and ceilings, ferrocement is not the answer for underground construction. On the other hand, curved shapes in concrete may be possible only with ferrocement methods.

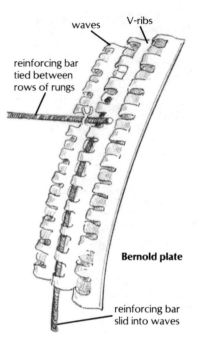

Figure 5–17: Bernold steel plates are good for arched-roof construction. The steel plates form the shell of a house as well as the concrete reinforcement. Shotcrete or gunite can be applied to both the interior and exterior surfaces.

The Armature

With cast-in-place construction, reinforced concrete is held in place by forms during the pouring and curing process. The reinforcing is strictly for added tensile strength and usually consists of fairly massive bars of steel placed where they will provide the greatest strength and advantage. In ferrocement construction, the steel bars and mesh provide strength and a support for the concrete while it sets. To support wet concrete, continuous welded-wire fabric is used rather than separate steel bars, although often both are used in house design. The main advantage of this special high-tensile-strength reinforcing mesh is that it allows a very thin, flexible, yet strong shell. While that is needed for boats, the much thicker construction needed for an underground house allows the use of multiple layers of less-expensive, lower-strength welded-wire mesh. One system of construction that is attractive for house building uses a sandwich of regular rebar between two layers of inexpensive 1-inch welded mesh. That, when properly engineered, allows strategic placement of high-strength steel in spots where strength is important, without the cost of continuous layers of high-strength mesh.

Even chicken wire (light galvanized wire netting) in combination with rebar reinforcing is used in some applications. The rebar provides the structural strength, while the chicken wire holds the concrete in place until it cures, reducing the likelihood of cracks in the finished panel. In boat hulls, it's common to have four layers of chicken wire on each side of the rebar. The use of many layers of mesh makes it harder to penetrate the resulting wall, and this can be a major problem in ferrocement construction if all the desired openings aren't formed in advance.

Photo 5-12: Bernold metal plates can provide the forming and reinforcing for barrel-vault construction. The Bernold plates can be placed quickly and usually at a lower cost than other forming methods.

A patented armature system that is particularly good for the barrel-vault style of construction (which has an arched roof) is United States Gypsum's Bernold sheets. As figure 5-17 shows, the sheets are steel plates formed into corrugated sections with loops of steel punched at fraction-of-an-inch intervals. The Bernold-sheet system was originally designed for tunnels, but it's practical for use in a variety of underground structures, including houses. The material is quite rigid and requires no additional reinforcement for most applications. Concrete can be sprayed onto both sides of the sheet or troweled into place.

Applying the Concrete

When building thin boat hulls, hand plastering is possible, although very labor intensive. Methods for hand application are essentially the same as for plastering, except that greater care has to be taken to fill all the space around the steel rods and the wire mesh. With the thicker panels that are needed for underground construction, placing the concrete by hand is not practical, both because it is impractical to adequately force concrete into a thick armature and because a much larger amount of concrete must be placed. Sprayed concrete, sometimes called *shotcrete, pumpcrete* or *gunite,* is a much more practical way to apply concrete. While all of these pressure-application systems force the mix through a flexible hose to the point of application, there are differences in the way the mix is created. In a pumpcrete or shotcrete application, the mix is made at the batch plant and then dumped into a pumping machine. With gunite, the cement and sand are mixed in the pump, and water is added by the "gunner" at the hose nozzle. Other pressure-application systems use other mixing systems.

With ferrocement construction, the gunite method, which allows the applicator to control the water content, is very useful. An experienced operator can tell how much water is needed in a mix by the way the mortar "rebounds." A mix that's too wet will have no rebound and will sag and tend to fall out of the mesh while a mix that's too dry will bounce off the dry steel mesh. Since the best mortar is made with a minimum amount of water, the operator has to adjust the water input to a minimum so that the cement sticks to the steel but does not flow. The gunite system provides this control, but it also means that the cured surface will not be smooth. But in underground structural panels, exterior finish is not important, and the superior strength of dry-mix gunite is an advantage. Gunite requires an experienced operator, however, and pumpcrete may be a better choice for most do-it-yourself builders, although it may be difficult for the inexperienced.

Whatever method of application is used, it is of paramount importance to fill all spaces around the steel bar and wire mesh with concrete. For thick application, it may be necessary to work from both sides of the armature or to provide some firm surface on the side opposite the pressure applicator to keep the pressurized concrete from deforming the back side of the armature. A person inside holding a 2-by-2-foot piece of plywood behind the point of

Photo 5–13: Bernold plates, originally used in Switzerland to construct tunnels, were first used in a residential application in the Stearns residence, which is made of two intersecting, circular arches, one 24 feet in diameter and the other 48 feet in diameter. Pictured here are the north and west sides of the house. Since the site is not suitable for passive solar, a water source heat pump interfaced with an active solar system and a woodstove with a water heating coil provide space heating and domestic hot water. The house is insulated on the exterior with 6 inches of extruded polystyrene insulation that tapers to 2 inches at the base.

application may be sufficient, although an interior form built in place provides a better interior finish but at much higher cost.

Finding the appropriate pressure-application equipment, particularly for gunite, may be difficult. If you live in an area where there are a lot of concrete swimming pools, contact the pool builders, since they often use gunite. Contractors who build commercial buildings may be a source of shotcrete or pumpcrete equipment. If you can't find pressure-application contractors or equipment, do not attempt an underground ferrocement house using hand-plastering methods unless you have literally dozens of workers. Even if you have the energy to move the tons of mortar from the mixer to the site and to place it, it is impossible to place the relatively dry concrete fast enough to prevent partial curing during placement, unless you have a large team. If you decide to try hand filling of a thick underground ferrocement shell, assume that you will need two workers for each yard of cement to be placed during each working day. Since an underground house may require over 100 yards of concrete, you can see the problem.

Spreading the plastering over many days is not a solution because you will have many cold joints that will reduce the strength of the shell and may become leak sites.

To take the best advantage of the good characteristics of ferrocement, use it only where curved surfaces are needed. A good example of this application is in a domed roof structure (see chapter 8 for an example). The walls are made with poured reinforced concrete and the ferrocement is used just for the dome. This is an especially good approach if an interior dome surface can be set in place to provide a backup for the armature and to function as an interior finish. Steel or fiberglass silo caps, wooden planking and geodesics are used in this way. Since the strength comes from the concrete applied over these domed surfaces, the material can be decorative rather than structural in nature. Although making a lightweight wooden dome requires a lot of skill and labor, it does not need to be expensive since it can be "planked" with nonstructural wood. With wood, don't forget to cover the dome with a plastic film before building the armature, or you may have concrete oozing through the joints.

Another place where ferrocement techniques can be put to good use is in exposed walls. Curves and free-form expression can use the thin-shell construction method used for boats as long as the primary structural strength is provided by columns and beams included in the wall at strategic points.

Summary of Ferrocement

By using reinforcing steel and wire mesh made of high-strength steel, a complete steel-reinforced concrete panel can be made without forms. Pressure application of the concrete ensures the complete filling of the wire-mesh armature, especially in panels over 1 inch thick. For structural walls that are flat, the cost of ferrocement is likely to be higher than for poured, reinforced concrete. For curved panels, however, ferrocement may be much cheaper. Structural analysis is very important, and engineers with design experience in loaded, curved, ferrocement panels may be hard to find, and also expensive when you do find them. Of all the shapes suited to ferrocement, domes are probably the most cost-effective, since they are constructed of thinner concrete and are relatively easy to engineer. Free-form, exposed walls that are not load-bearing are also a good application. Insulation installation, however, is a problem on curved surfaces. Interior (rather than exterior) insulation is a good approach for exposed walls, and layers of rigid foam insulation are practical choices for flat surfaces.

While ferrocement offers certain advantages to the owner-builder, structural surfaces of ferrocement may be more costly than cast-in-place methods.

Construction Using Concrete Block

Of all the methods that use concrete for the shell of an underground house, the only one readily available to the owner-builder who wants to do it all is construction using concrete masonry units commonly known as concrete blocks. Of course, there are concrete blocks and there are concrete blocks. Some are made with cinders or fly ash, some are made with regular rock aggregate and

Concrete Blocks

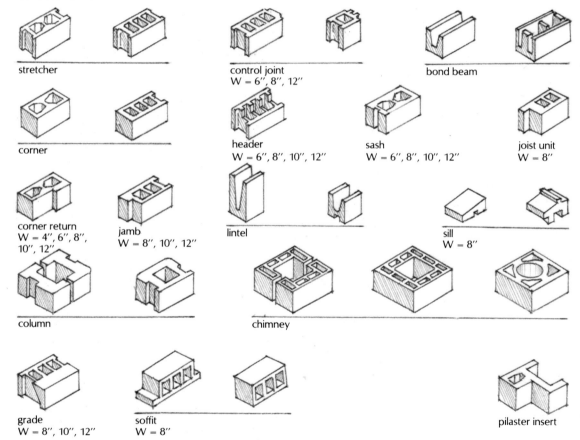

Figure 5–18: This illustration shows typical sizes and shapes of concrete blocks. These concrete blocks are available in the following sizes: 4 or 8 inches in height; 8, 12 or 16 inches in length; 2, 3, 4, 6, 8, 10 or 12 inches in width. Some blocks are available only in certain widths, which are indicated. Column, chimney and pilaster insert blocks are available in special sizes.

SOURCE: Redrawn from C. G. Ramsey and H. R. Sleeper, *Architectural Graphic Standards*, 6th ed., copyright © 1970, with permission of John Wiley & Sons, New York, N.Y.

*Figure 5–19: Bond beams in block
walls provide extra strength
against both vertical and horizontal
loading of a concrete block wall.*

SOURCE: Ronald Smith, *Principles
and Practices of Light Construction,* 2d ed. (Englewood Cliffs,
N.J.: Prentice-Hall, 1970).

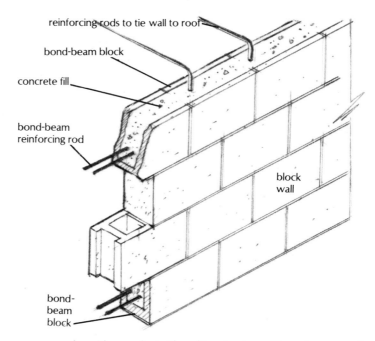

some are even made with sawdust. The ultimate strength really comes from steel
reinforcement and concrete that is poured into the hollow cores of the blocks
rather than from the blocks themselves. Even with high-density blocks, the fact
that low-strength mortar joints separate them means that unfilled block walls
don't have enough strength to hold back the 7 to 10 feet of earth around an
earth-sheltered house. The ability of an unfilled block wall to withstand pressures
from the adjacent earth (horizontal strength) is less than that of an unreinforced
concrete wall about half as thick. To be structurally adequate, a block wall needs
more than just mortar to tie the blocks together.

Reinforcing Concrete Block

Concrete blocks are typically made in widths from 2 to 12 inches, but only 8-
or 12-inch units should be considered for a wall that has any side loading, such as
in an underground exterior wall. The block cores must carry both the reinforcing
steel and the poured concrete fill (*grout*) to have the necessary horizontal
strength. Interior walls that have only vertical loading (from the roof) can be made
from 6-inch block, but the 4-inch unit has such a small core opening that it cannot
effectively be reinforced, so 4-inch block is not usually used for full-height
structural walls.

There are a number of different block shapes, some purely decorative and
some with special shapes to allow enhanced strength when they're filled. Most of
the decorative shapes are intended to give an attractive texture to the exterior of
a structure, and they are not of much use in an underground house unless the

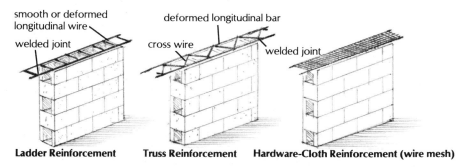

smooth or deformed
longitudinal wire

deformed longitudinal bar

welded joint

cross wire

welded joint

Ladder Reinforcement   **Truss Reinforcement**   **Hardware-Cloth Reinforcement (wire mesh)**

Figure 5-20: Various types of joint-reinforcing materials provide horizontal strengthening of block walls. Ladder reinforcement and truss reinforcement use wire, while hardware-cloth reinforcement uses a wire mesh. The reinforcing wire or wire mesh is placed between horizontal layers of concrete block, thereby tying the block together and preventing cracks that may occur from enlarging. The reinforcing material is thin so that it fits into the mortar between the blocks, and while it provides reinforcement, it is not intended to be as strong as the reinforcement provided by steel rebar.

textured surface is desired for the finished interior. One special block style, called the bond-beam block, is important in underground construction. With the bond-beam block, rebar can be run both horizontally and vertically with plenty of space left for the concrete fill (see figure 5-19). With a row of filled bond-beam blocks in place with two continuous reinforcing bars in the core, a structural beam or column is created. Most aboveground structures built from block include a bond beam at the top of a wall or at every floor level in multi-story buildings. A subsurface wall will have considerably higher horizontal stress and may need a central horizontal bond beam as well as one at the top of the wall.

A common block construction system for underground houses uses blocks that measure 12 by 8 by 16 inches (12 inches thick, 8 inches high and 16 inches long) in combination with ½-inch (number 4) rebar vertically placed in each cell about 1 inch from the inner wall of the block and centered left and right in the block cell. Horizontal reinforcing includes two continuous number 4 bars in a bond beam along the top and another pair of number 4 bars in a bond beam halfway up the wall. Also, steel-wire ladder reinforcing is laid between every other row of blocks to increase horizontal strength (see figure 5-20). Of course, all voids are filled with quality concrete, and the bars are arranged so that the concrete will completely cover them. Backfilling should not be done until the concrete has set at least a month or more. This particular structural design may be inadequate in some sites, and a structural engineer should always make the final judgment as to amount and placement of reinforcing.

When ordering the grout (the concrete to be used to fill the voids in the block), be sure to specify an aggregate that is sized for bond beams. If you do not, you will probably get concrete that will have aggregate too large to enter the bond-beam cavity between blocks, and the strength of the wall could be

Figure 5-21: In underground houses with grouted block walls and either a poured-in-place roof or a roof of precast panels with a poured concrete cover slab, the vertical reinforcing in the walls should extend into the roof.

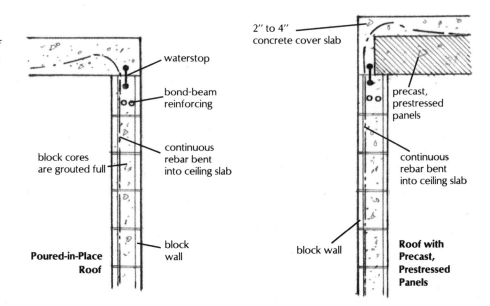

Poured-in-Place Roof

waterstop

bond-beam reinforcing

block cores are grouted full

continuous rebar bent into ceiling slab

block wall

2" to 4" concrete cover slab

precast, prestressed panels

block wall

continuous rebar bent into ceiling slab

Roof with Precast, Prestressed Panels

significantly lowered. Although it is common practice to fill each column of block cells individually by simply backing the premix truck to the wall and running the concrete into the cores via the truck chute, I suggest that all the cores be filled first to the level of the first bond beam, then as evenly as practical until the entire wall is grouted. If that is not done, the horizontal movement through the bond-beam cavity will cause separation of the aggregate, and nonuniform concrete of reduced strength will result. If there is no intermediate-height bond beam and only one at the top of the wall, then that precaution will not be necessary.

If a poured-concrete roof is to be placed on top of the block wall, the engineering design will usually include leaving the vertical rebar extending at least 18 inches beyond the top of the block. Before the roof is poured, those bars are bent into the line of the roof and tied to the roof reinforcing to provide a strong wall/roof attachment. If the roof is to be made with precast, prestressed panels, the rebar will usually be cut an inch or so below the top of the block so that the grout can be screed (scraped while wet) flush with the block top. This gives the roof panels a flat base for support.

As figure 5-21 shows, structural systems that include a poured-concrete pad over the precast, prestressed panels may be arranged so that continuous reinforcing passes between the wall and the roof. To do that, the precast panels are ordered a foot short so that when they are centered over the span, 6 inches of the 12-inch blocks are not covered. The vertical core reinforcing extends above the block in that gap and is bent over so it will be cast into the surface slab that is poured over the precast panels. The openings in the precast panel ends should be blocked with building paper or roofing felt or plugged with wads of newspaper to keep the pour from filling the cavities and increasing the weight of the roof with no appreciable increase in strength. With this method, the precast panels provide the primary strength and a form base for the poured slab, which increases wall strength and positively ties the roof to the walls. With this system, a complete shell can be built without the need for forms except along the edge of the roof cover slab. Since framework is likely to be the main stumbling block for the owner-builder, there is a clear advantage in this approach.

If you've seen it done, you may have thought that block laying *looks* easy, but if you're new at it, don't bite off more than you can chew. Many amateur masons have done excellent work, but it takes practice to lay up a perfectly vertical wall in a straight line, and the place to practice is not with heavy 12-inch blocks on a wall that is 80 feet long. If you are going to do the block laying yourself, the best way to learn about it is to apprentice yourself to a block layer. Barring that, first build something of more modest size with lighter blocks, even if you tear it down later on. If you do that, be sure you build something that has a straight stretch with right-angle corners at each end. You will learn very quickly why masons work from the corners in and not the center out.

Other Block Systems

Another system is the dry-stacked fiberglass-reinforced or *surface-bonded* block wall. It has been used for years, particularly by amateur masons. In this method, the blocks are stacked without mortar, and then a coat of fiberglass-reinforced epoxy resin is troweled or sprayed over the exterior surface of the blocks. Ignoring structural considerations for the moment, there are some problems with this system. The one that causes the most trouble is the fact that block sizes are on 8-inch modules including the thickness of the mortar joints. When 12 8-inch blocks are laid up with mortar, the result is a wall 8 feet high. When stacked dry, however, the result is several inches shy of 8 feet. While that is not necessarily a problem in a solid wall, the frames for doors and windows that are intended to fit in block walls will not fit if mortar has been left out, so if you are making interior walls from block and the exterior walls are laid up without mortar, you have the choice of laying the interior walls with mortar but having a mess at the ceiling where too small a space is left for a full block, or laying up without mortar and not having standard door openings.

Of course, if you use wood- or steel-studded interior walls, then the only problem you will have is to trim standard 4 × 8-foot wall coverings (e.g., drywall) to fit the nonstandard wall height.

The adequacy of the dry-stacked wall in horizontal strength is of great concern. While such walls have been used in retaining-wall applications and even in underground buildings, there are many different coating materials available, and each one has its own capabilities. Before committing to such a system, consult the manufacturer of the product — asking specifically what he will do for you if the wall fails — and consult a professional engineer for a second opinion.

Besides the common 8 × 8 × 16-inch and 12 × 8 × 16-inch block, some block manufacturers make special blocks for special applications, such as silo blocks. If you are in dairy country, curved blocks for silo construction may be available for making curved entryways, rounded corners, stairwells, atrium cores and all sorts of interesting curved shapes. With the curve facing into the load, the resulting arch is stronger than the flat surface made from square block.

How to Increase Strength of Block Interior Walls

Interior partition walls of block add dramatically to the effective strength of the exterior walls because of the buttressing effect of the partitions. Interior walls intended to carry some of the stress from the exterior walls should be grouted full and built on a foundation slab just like the exterior walls. They may, however, be as thin as 6 inches, because the load is along the long wall axis, not across the thickness. Steel reinforcing may be needed only every third block (4 feet) instead of two per block as in the outside wall. Horizontal reinforcing may not be needed either, although a bond beam at the top is considered good practice. The foundation for the interior walls should be continuous (including reinforcing) with the outer wall foundation, and if the main foundation and exterior walls are not joined to the floor slab, the interior walls should be left free-floating also.

Lightweight and Insulating Block

Some block is made with lightweight aggregate such as cinders and processed fly ash. Others include insulating materials such as sawdust or even polystyrene beads. As long as they can support themselves until the grout fill has cured, these blocks can be used with only a slight loss of structural strength. Their performance as insulation usually leaves a lot to be desired, however, and if they are found to be more expensive than common blocks, an extra ½ inch of foam-board insulation will probably be cheaper and more effective as insulation. The only valid reason I can find for using lightweight blocks is that they are lightweight. Laying regular concrete blocks — especially 12-inch blocks — can tire a decathlon winner in a few hours, and while lightweight blocks may be only 30 percent lighter, that is a significant difference. Whatever kind of block you do decide to use, be sure that any engineering advice you get is based on that specific block. While you will not have strength problems if you ask for detailing using lightweight block and then decide to switch to common block, the opposite is not true.

Block–Roof Construction

While concrete-block roofs are not at all common, they are possible. They are made by laying block (often 4 × 8 × 16-inch) between supporting stringers of steel or prestressed concrete. In that configuration, the cores are usually left open, and a reinforced slab is poured over the whole assembly. This roof system allows construction without the difficulty and expense of roof forms, but it does require finishing on the ceiling side, whereas precast, prestressed panels, which offer the same advantages, are often left unfinished as a ceiling.

Earth-Contact Houses

The emphasis in this book is on designs in which the earth covers the roof as well as much of the vertical wall area, but the *earth-contact* house is another design option. The outstanding feature of this type is the use of earth berms, which usually cover three walls either partially or completely up to the roof. While earth-contact houses and underground houses have similarities, the differences are significant. It is certainly possible to design an earth-contact house that provides a favorable combination of aboveground and underground features, but it is also easy to end up with a structure that has few of the advantages of either and the disadvantages of both.

As a rule, the most commonly touted advantage of earth-contact housing is its energy efficiency. Unfortunately, many of the earth-contact houses that have been built could have achieved similar or even better energy efficiency by using conventional aboveground construction with additional insulation. The difficulty usually lies in the parts of the walls that are within a foot of the surface and that extend above the surface. When wall construction is of concrete, that upper wall portion will need at least 4 inches of foam insulation to equal the insulation found in good-quality frame construction. At a foot or more below grade, the need for insulation will be reduced according to the same rules that apply to underground houses. The basic problem is one of protection of the insulation. It is easy enough to put 4 inches of foam against the outside of the wall, but sunlight, lawn mowers and even rodents will cause damage to the material and gradually reduce its insulating value. Besides, the appearance of bare foam is usually not suitable for the exposed side of a house. More often than not, the "solution" is to use 2 or just 1 inch of foam to make the fastening of siding over the foam practical. Naturally, the result is a wall section that it not nearly as well insulated as even the cheapest tract house. There are, however, several methods for keeping a high insulation value over the entire wall surface. The most conventional is to insulate on the inside. The masonry wall is furred out with 2 × 2s or 2 × 4s, and the cavities are insulated either with foam (to keep the frame wall thin) or with fiberglass (which is a lot less expensive). This furred-out surface is also very convenient for making wiring and plumbing runs that would otherwise be difficult or expensive to install in a concrete wall. The clear disadvantage is that the mass of the concrete walls is not available for heat storage. A second method is to install the thick foam on the outside of the concrete and then cover it with one of the proprietary spray or trowel-on coatings made specifically for protecting foam insulation placed on the outside of concrete. For this to be satisfactory, the foam must be firmly glued or otherwise fastened to the concrete. Steel Z-strips along the foam boundaries combined with glue strips or spots provide a reasonable system of attachment (see figure 5–22). The glue must be compatible with the foam material. For example, if polystyrene is the insulation, then any panel cement including a petroleum-based solvent or binder will dissolve the foam. Dow Chemical Company, the manufacturer of Styrofoam, has an adhesive called Styrofoam Brand Insulation Mastic #11, which is made for the purpose. Adhesive clips are also available from the Stic-Klip Division, Ayer, Massachusetts, for holding the edges of insulation panels in place. As figure 5–22 shows, regular stucco can be used for a protective and decorative covering, although the cost may be rather high in some localities. Latex-modified stucco with fiberglass-mesh reinforcement, latex-modified cement mortar, cement-asbestos board, treated plywood or rigid vinyl sheeting (normally sold as soffit

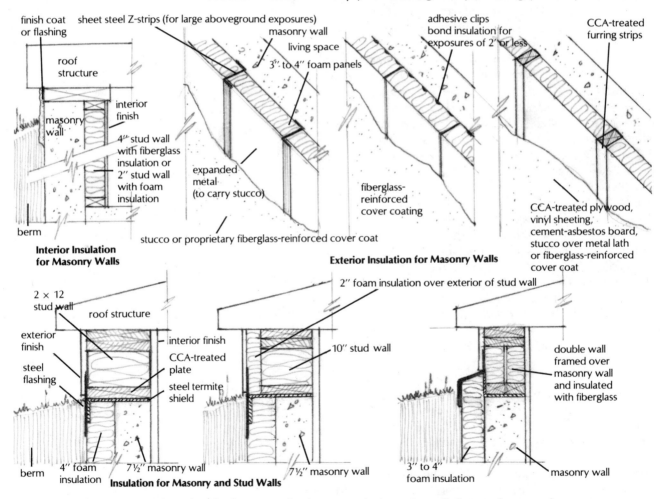

Figure 5–22: There are a number of methods to use to insulate properly the entire wall of an earth-contact house.

material or siding) can be used if clips, Z-strips or treated wood strips are placed at the joints of the insulation when more than a few feet of insulation surface is out of the ground.

Another approach to the problem is to stop the concrete a foot or so above grade and use frame construction for the remainder of the wall. That is particularly useful where large surfaces extend above the berm, as may be the case with the sidewalls of houses where only the north wall is completely bermed. The primary design difficulty is to produce a continuous surface on both the inside and outside of the wall. The effective thickness of the concrete section is 12 inches, with 8 inches in concrete and 4 in insulation. To keep surface alignments both inside and outside, a frame wall that is effectively 12 inches thick is needed. While that provides an extraordinary opportunity for insulation (R-38 with fiberglass), it may be expensive. For some designs, built-in cabinets can be fitted above the concrete, but an 8-inch frame wall is still required with 4 inches providing the structural support and 4 providing the extension necessary to keep the outside aligned with the insulation. A final variant is to use a thinner wood wall, extend it only 2 inches beyond the concrete and put 2 inches of foam over the wood. The siding can be nailed through 2 inches, but having more than that can present a fastening problem.

In many cases, the most cost-effective construction method is to use lumber treated with chromated copper arsenate and build according to All-Weather Wood Foundation methods discussed later in this chapter. This all-frame type of construction allows the insulation to be placed between the studs; there are no material changes with corresponding problems at the transition between below grade and above grade; and costs are usually lower than those for comparable masonry construction. The thermal mass of wood is not going to provide significant heat storage, though, so there may be a need for additional water or masonry mass, especially with a solar-heated design.

The same waterproofing designs should be used for earth-contact houses as for underground houses. Since one of the keys to success in waterproofing is keeping water away from the wall surfaces, roof runoff should be collected in gutters and piped at least 10 feet from the house walls. Roof overhangs extending several feet over the berm help keep the area near the wall dry.

Untreated wood should be kept at least 8 inches from the earth. In particular, overhangs should have an 8-inch or larger aboveground clearance so that there can be adequate air movement under the overhang. If these precautions are not observed, high moisture levels may be present, and dry-rot damage may result. Also, if continuous foam insulation on the exterior of the wall is used, termite shields should be installed to prevent insects from crawling into the house from behind the insulation. Figure 5-22 shows a typical shield in place.

Quite frankly, a very significant part of the energy-conservation advantages that are provided by earth-contact construction result from reduced window

Photo 5-14: Access to the roof of an earth-contact house may be so easy that it becomes a nuisance if the roof "attracts" children and animals. Each site will determine how significant this might be.

areas and greatly reduced infiltration losses. Many windows are placed in aboveground houses often because they look good from the outside rather than because they are needed by the occupants. Large windowless walls may not be aesthetically acceptable, but landscaped earth berms are. In such cases, earth berming can provide better energy performance than even superinsulated designs. In climates where cooling is the primary energy user, berming on the east and west and, to a lesser extent, the south will be most useful. Once again, the elimination of windows on the east and west will have a greater effect on the cooling loads than just about anything else that can be done.

One disadvantage peculiar to a house bermed with soil up to the roof is the ease of access to the roof by people and animals. The roof of one earth-contact house was once badly damaged by a wandering cow. Traffic by dogs, children and goats is a definite hazard. A relatively steep roof slope is one solution, and fencing is another. You will have to be the judge of how significant the problem is in your location, but roof maintenance can be an unplanned but continuing expense if the possibility is ignored.

Summary of Earth-Contact Houses

If close attention is paid to the continuity and quality of the wall insulation within a foot of grade and above grade, earth berming can be as energy efficient as underground construction. The cost of a conventional roof is usually less than one designed to be covered with earth, so there can also be a cost savings. You should, however, treat the underground portion just as you would if the whole structure were below the surface. Scrimping on structure, waterproofing or insulation is not worth the risk. If you cannot find a builder to do the job right, consider a superinsulated house instead of an earth-contact design.

Be sure roof runoff is carried well away from the structure. Keep untreated wood at least 8 inches above the soil line, and allow adequate ventiliation under roof overhangs. Don't forget the ease of access to the roof surface that comes from a high berm.

Some advantages of bermed construction in contrast to underground construction are that it's usually cheaper; it's easier to find competent contractors, since conventional roof construction methods can be used; egress from isolated rooms is sometimes easier; daylighting through small windows between the top of the soil and the roof is possible; simple construction methods for cathedral ceilings can be used; and the more conventional designs mean fewer problems with building codes.

Some of the disadvantages are that there is increased maintenance, since the roof is exposed to the weather; bermed houses are difficult to insulate; they may require a stronger wall structure, since there is no rigid tie between walls (provided by the reinforced concrete roof in an underground house); bermed houses are less fireproof; and there's more potential for infiltration heat losses.

Wood Construction

Although many earth-contact houses have been built, with earth berms up to the roof line and with a conventional wood roof above grade, few earth-sheltered houses have wood for the vertical walls. It is possible, however, to have all-wood construction in an earth-sheltered design, and in many cases it can be less expensive than similar designs using concrete.

For the last 50 years or so, building systems have used treated wood placed below grade. Recently, a below-grade structure built with treated wood was disassembled after 40 years, and it showed no signs of deterioration. In the last 20 years, an estimated 15,000 basements have been constructed using what has come to be called the All-Weather Wood Foundation (AWWF).

The preservative used in treated wood is a water solution of chromated copper arsenate (CCA). All wood to be used below grade is treated with the CCA-water solution to a specified minimum retention level. The end use of the wood determines what preservative retention level is required. Levels of 0.25, 0.4 and 0.6 pounds of CCA per cubic foot of treated wood are standard values, and a 0.6-pound-per-cubic-foot retention is appropriate for permanent underground construction. After treatment, the wood has a greenish coloration that

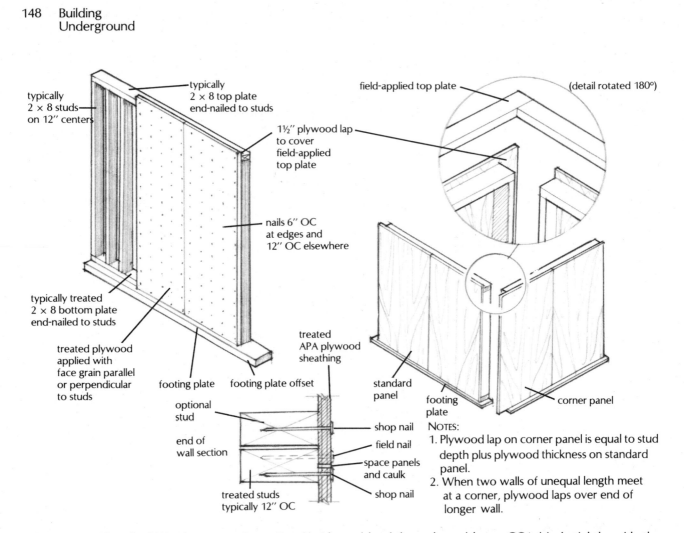

typically
2 × 8 studs
on 12" centers

typically
2 × 8 top plate
end-nailed to studs

field-applied top plate

(detail rotated 180°)

1½" plywood lap
to cover
field-applied
top plate

nails 6" OC
at edges and
12" OC elsewhere

typically treated
2 × 8 bottom plate
end-nailed to studs

treated plywood
applied with
face grain parallel
or perpendicular
to studs

footing plate

footing plate offset

treated
APA plywood
sheathing

standard
panel

footing
plate

corner panel

optional
stud

end of
wall section

treated studs
typically 12" OC

shop nail

field nail

space panels
and caulk

shop nail

NOTES:
1. Plywood lap on corner panel is equal to stud
   depth plus plywood thickness on standard
   panel.
2. When two walls of unequal length meet
   at a corner, plywood laps over end of
   longer wall.

Figure 5–23: Although All-Weather Wood Foundation (AWWF) systems are built using conventional framing methods, the structural requirements are much greater for underground systems. Larger framing members and attention to several details are necessary.

SOURCE: Redrawn and adapted from *APA Design/Construction Guide: All-Weather Wood Foundation,* copyright © 1978, with the permission of American Plywood Association, Tacoma, Wash.

can be painted with no bleed-through problems. CCA binds tightly with the wood's cellulose fibers, and even if the treated wood is eaten, no toxic reaction will occur. Likewise, there is no bleeding of the preservative into the surrounding soil, so even growing beds can be made of CCA-treated lumber, whereas creosote or pentachlorophenol treatment may cause problems.

While treating wood with CCA is not practical for the do-it-yourselfer because pressure application is needed for adequate penetration of the preservative, CCA-treated wood is readily available in all structural sizes and also in the various thicknesses of plywood commonly used for building construction.

AWWF Construction Methods

Conventional framing methods are used in All-Weather Wood Foundation systems. The stud sizes and spacing, however, will vary with the soil type and the location of load-bearing interior walls. In basement applications, a typical design uses 2 × 6-inch studs on 12-inch centers sheathed with ½-inch plywood. But, as with concrete building systems, basement construction and underground-house construction methods are not the same because of the increased loads and the more stringent appearance requirements of a finished living space.

For underground construction, the National Forest Products Association recommends placing the base of the treated-wood wall on gravel rather than on a conventional footing or directly on the undisturbed soil. This ensures good water drainage under the floor. With the high footing loadings likely to be encountered in a fully underground house, concrete footings should be considered with under-the-floor drain tiles added for under-the-floor water removal.

Roof construction can also be done with CCA wood, but careful analysis of the structural requirements must be made, and adequate rafter size and spacing is important. The roof should be able to carry a load of at least 120 pounds per square foot for every foot of earth cover that is planned. The lumber chosen for the roof structure must be of as good a grade and of the same species as assumed in the strutural calculations. Pine is a lot weaker than Douglas fir, and number 2 grade is weaker than number 1 structural. It may turn out that if you have over a foot of earth on the roof, you will have so much wood in the roof structure that any cost savings over concrete may well be lost, so it's a good idea to make careful comparisons. My personal recommendation is that if you choose all-weather-wood walls, you should build a conventional, exposed roof on top rather than one that is fully underground.

Because of the potential for corrosion, all nails, fasteners, clips and braces have to be corrosion resistant. Stainless steel is preferred as a basic material, although silicon-bronze and copper are acceptable where strength requirements are not so stringent. Hot-dipped galvanized steel can be used if it's behind the waterproofing, but zinc electroplated fasteners are not recommended. Although many people believe that aluminum does not corrode, it actually corrodes very rapidly below grade and will literally dissolve in just a few months in some soils. Thus, aluminum should never be used either for flashing or for fasteners that are in contact with the soil.

Advantages of AWWF Construction

All-Weather Wood Foundation systems are often lower in cost than concrete of similar structural capability. That is true particularly of walls, but roof structures underground have to be so massive when wood is used that there may be no cost advantage. Insulating the wood-frame underground house is much easier and cheaper because standard fiberglass-batt insulation can be placed between studs and rafters. Of course, a lot of cutting and fitting may have to be done, since the standard widths of fiberglass batts are not likely to be usable with the close stud/rafter spacings needed in underground construction. Exterior rigid foam insulation can be used as well, but the advantages of exterior insulation are largely lost with wood construction, since it has so little mass.

With wood, a variety of interior surfacing materials are available. Drywall, paneling and everything that can be installed in an above-grade house are possible in the AWWF house.

Conventional methods of wiring and plumbing are also possible. No conduit is necessary, and all wiring can be in the frame of the house. In fact, the only unusual aspects of underground wood construction are the close spacing of structural lumber (studs, beams and rafters) and the need for quality waterproofing. For that reason, it may be easier to find contracting services for the underground wood house than the concrete house.

Weather is less of a problem in the construction of a wood house. Freezing weather, rains and unusually hot weather may halt the construction of a concrete shell, but not wood construction. Concrete also must cure for several days, even weeks in some cases, but wood is, of course, finished as soon as the erection is completed. Highly expansive clay soils may cause bowing of wood walls when the ground is unusually wet, but the same pressures might crack concrete. However, such a soil condition would call for extra drainage measures that would eliminate the problem no matter what the shell was made of. One advantage often cited for wood walls over concrete is less dampness in the space. That is not a function of the material but rather of how the material is insulated. In a basement where there is no insulation over the concrete, condensation can occur, especially in northern climates, and dampness may be a problem. The wood wall, however, is less likely to have condensation problems when uninsulated.

The greatest advantage of AWWF construction is for owner-builders, since construction involves standard carpentry skills and tools, manageable structural requirements, conventional roof framing and the elimination of masonry insulation problems where walls extend above grade. These factors, along with possible cost savings, make all-wood construction worth consideration by owner-builders.

There are, however, a couple of drawbacks to AWWF construction that have to be considered in the design stage. The low mass of wood construction may greatly reduce the performance of a passive solar space–heating system. With less mass available for absorbing surplus solar heat, there will be wider daily variations in indoor temperature. The radiative effect of warm masonry walls is lost, and more elaborate heat distribution systems may be needed. One solution is to include more thermal mass in floors and interior partitions.

Wood also burns. Fire damage of the main structure is unlikely in a concrete house, but that certainly isn't the case with AWWF construction. Careful consideration of the placement of multiple exits from each room should be an important part of the design process. The possibility of fire can be significantly reduced by covering the wood with one-hour fire-rated drywall. This "X"-rated ⅝-inch-thick gypsum board is designed to keep flames from burning through the

Photo 5–15: With the exception of a concrete slab floor and concrete footings for the retaining walls, the Zanetto house is constructed entirely of wood. The below-grade walls are of AWWF construction methods, and the roof uses 2-inch joist framing. The east and west walls are totally bermed, the north wall is partially bermed, and the roof has 8 inches of soil cover.

drywall for at least one hour. Commercial buildings often have two layers of fire-rated drywall for extra protection. If you have an earth-covered roof, one place to be particularly cautious is at the chimney. The earth surrounding the flue pipe keeps it from cooling off, and surface temperatures will be a lot warmer than if the penetration were exposed to open air. Even if temperatures are not high enough to be a fire hazard, waterproofing at that point may be difficult. To keep that from being a problem, you can run the chimney pipe through a larger pipe to create a vented air gap between the two. The larger pipe is sealed to the roof, and the chimney pipe itself is not in contact with the waterproofing or the wood structure.

For more complete details on the use of CCA-treated lumber in the AWWF construction system, write to the National Forest Products Association, 1619 Massachusetts Avenue NW, Washington, DC 20036. They have complete manuals for the use of wood in basement applications. Remember, however, that basements and underground structures do differ, and you'll still need engineering services for your specific site and design.

Waterproofing and Insulation

There is no great mystery to making an underground house watertight and thermally comfortable. Proper surface and subsurface drainage of the house site and the proper installation of waterproofing materials such as paint-on materials, waterproofing membranes and montmorillonite clays eliminate water problems in underground houses. Since the soil surrounding an underground house isolates it only from the extremes of aboveground cold and heat, insulation is nearly always necessary to make an underground house energy efficient and thermally comfortable. But insulation requirements vary according to geographical area and type of house, and some types of insulation are preferred over others. This chapter explains basic waterproofing techniques and tells how to make the best use of insulation.

Waterproofing

People contemplating underground construction worry about waterproofing more than anything else. I suspect that this is partly due to stories of leaky basements. Charles Wing, well known for his books on house building and for Cornerstones, his trend-setting owner-builder school, actually considers basements as wells that sometimes run dry. My own experience with basements has also been rather poor; perhaps yours has too. Yet, I have visited only one underground house that has had problems with leaks of any significance, and that was easily traced to the concrete contractor having left out waterstops between the walls and the floor. What is the difference? Why the abominable record for basements and the excellent one for underground houses? I think there are two basic reasons. First, a basement rarely gets more than a single coat of an asphalt emulsion for its "waterproofing," while underground houses are usually swaddled in one or more membranes, coatings or panels of high-quality, impervious materials. Second, drainage is always well considered in underground-house design. The great majority of underground houses have footing drains, and many have drains under the floor as well. Not only do basements rarely have perimeter drains (that work), but surface drainage patterns are not usually considered in the design and construction of a house. When it rains, water pouring off the roof is not always directed away from the house. Add to this the facts that cracks in basement walls and floors are common and that there is usually no water barrier between the walls and the floor, and I'm amazed that flooding is so infrequent!

While I have to agree with Charles Wing in his general condemnation of basements, the condemnation is really of common, often inadequate, construction practices. It is not particularly difficult to have a waterproof basement or underground building if the builder is willing to pay attention to a few details.

Drainage

If Wing's Law says that a basement is a well waiting to fill, then Wade's Law says that waterproofing works best where there is no water. By arranging land contours, drainage tiles and the content of backfill surrounding the underground house, water can effectively be kept from contact with the building. No matter how fine a job you do in applying waterproofing, if the house is immersed in water for extended periods, leaks are inevitable. Even the best boats need bilge pumps. As I have emphasized elsewhere, good drainage is the basis for good waterproofing, and that starts with surface contouring. In most parts of the

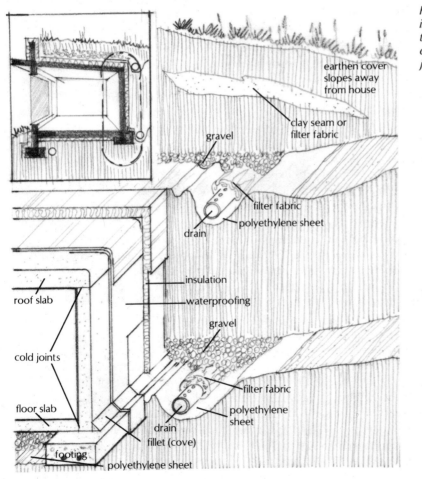

Figure 6-1: Preventing water from ever reaching an underground house is the most important part of waterproofing, and there are many details that comprise the total waterproofing job.

earthen cover
slopes away
from house

clay seam or
filter fabric

gravel

filter fabric

polyethylene sheet

drain

insulation

waterproofing

gravel

roof slab

filter fabric

cold joints

polyethylene
sheet

floor slab

drain

fillet (cove)

footing

polyethylene sheet

country, a heavy rain results in lots of surface water rapidly moving to a lower level. If your house occupies that lower level, you can expect problems. Moving water needs to be diverted away from the walls by means of drainage channels and grading of the earth around the house. The path of running surface water should be toward a spot at least 10 feet beyond the boundaries of the underground house. Under no circumstances should standing water be allowed near or on the house. Standing water usually implies saturated earth, which means trouble.

In addition to surface drainage, subsurface drainage needs to be considered in underground-house designs. Perforated pipe or tiles placed at or below the level of the footings are very important. These, when properly installed, will drain away water that passes through the earth after percolating down from the surface. Proper installation includes several important details. First, the tiles should slope toward an above-grade outlet. This means that somewhere within a reasonable radius of the house, for example, 100 feet, there must be a place where the subsurface water-collection system can drain into the open air. That may be a storm sewer specifically designed to rid an area of surface water after a rain or, where permitted, a sanitary sewer. Most often, however, it means a place where the surface of the ground is several inches lower than the house footings. If such a spot is not available on your proposed site, consider very carefully how important it is to you to build an underground house at that particular place. If energy conservation is your only reason for going underground and if you insist on building at that site, please consider an aboveground solar house or a superinsulated structure rather than one below grade. While you *can* do a good enough job of waterproofing the house surfaces or you *can* use a sump pump to drain the water collected by footing drains, both approaches are much riskier than well-placed gravity drains. The waterproofing around a house that is immersed in saturated soil must be a whole lot better than waterproofing around a house on a well-drained site. If you have ever been in the attic of an older house that has wooden shingles, you will se a surprisingly large number of cracks and holes penetrating the roof; yet, it doesn't leak except in extraordinarily heavy rains. That same roof would be useless in holding back water that had pooled on its surface. The same is true underground. If the water is pooled against the house under hydrostatic pressure — which is the case following a heavy rain if there are no footing drains — the waterproofing has to be a lot better than if the water is draining toward the footings.

One house I visited recently uses a liquid asphalt, tar-impregnated-felt waterproofing system, which I would never recommend for an underground house, yet the occupants have never had a leak. Why? The house is really built aboveground and covered with sandy soil that drains well, and there is a good network of footing drains. That particular house probably would not leak even if the bare concrete were left exposed.

When installing subsurface drains, remember that you are trying to keep water away from all parts of the house surface. To do that, water must have an easy path to the drain at all levels from the surface down to the footing. The most common method is to fill the space over the drain pipe with a very porous material such as gravel. That allows the water to percolate down to the drain rapidly enough so that accumulation against the walls of the house is unlikely. If the soil itself is sandy and water passes through it freely, then the gravel may be unnecessary except to surround the tile. The tile should be buried in gravel larger than the openings into the drain pipe to keep from clogging the pipe and to keep surrounding soil from washing out through the drain and creating a cavity around the drain, which will ultimately collapse, causing surface subsidence and probable drain clogging. To keep the gravel from mixing with the surrounding soil, it is often separated from the dirt with a layer of asphalt-impregnated felt (roofing paper). Of course, the water doesn't pass through the felt, but at laps and edges, there is plenty of opportunity for drainage.

In soil with high clay content (low permeability), the gravel layer should extend to the roof level. It does not have to be vertical, however. A sloping layer following the excavation contour is fine, as long as it is at least 6 inches thick and extends from the footing to near the surface. Beware, however, that the gravel layer does not drain surface water as well as subsurface water. Drainage tile has limited capacity, and if the gravel extends too near to the surface under a watercourse, it may actually be directing water to the house footings. In silt or fine sand, a layer of asphalt-impregnated roofing felt (so-called 15-pound felt) should separate the gravel channel from the soil to keep silt from filling the spaces between the rocks and slowing down water flow. The water will penetrate the loosely laid felt, but little silt will be carried into the gravel itself.

An interesting alternative to the gravel collection system is a material called *Enkadrain*. It is a polyester fabric with a mat of springy nylon filaments arranged to provide a free path for water down a wall surface. It is installed with the nylon mat against the wall exterior — over the insulation or the waterproofing, whichever is next to the soil. After backfilling, the nylon mat holds the polyester fabric away from the wall surface, providing a free path for water-to-footing drains. The idea is the same as with a gravel layer: If there is a free path to the drain tile, then the water is not under hydrostatic pressure, and waterproofing is less of a problem. The materials are not biodegradable, and the filter fabric is designed to allow free water movement but to block silt and fine soil particles, so it should work fine. However, it is not cheap. Where the homeowner is able to provide the necessary labor, the gravel-layer method is much less costly. On the other hand, hand placement of several yards of gravel by a contractor is not cheap either, so cost is a relative factor. The other potential problem is its use with erodible waterproofing materials such as bentonite. The water path is confined to a fraction of an inch beyond the surface of the wall, so water velocities are considerable. Running water will destroy clay-based waterproofing by eroding it away, but if the waterproofing is covered by rigid insulation and then by

Enkadrain or a similar material, there should be no erosion. Membrane types of waterproofing are not eroded by running water and should cause no problems.

Waterproofing Materials

After surface contouring and perimeter drainage, the primary line of defense against water entry is the building material itself. If it is concrete, low-water/rich-cement mixes and waterstops at all cold joints will provide a shell that is intrinsically waterproof. The second line of defense is the waterproofing materials placed over the outside of the shell. There are three categories of waterproofing materials: liquid-applied membranes (paint-on and cementitious materials), performed membranes (roll goods and sheet membranes) and montmorillonite clays.

Liquid-Applied Membranes

Liquid-applied materials include polyurethanes, portland-cement coatings, asphalt-based coatings and rubber-based paints. All suffer from one basic shortcoming: They have a tight bond to the surface, and if a crack occurs or one part of the surface moves in relation to another, even slightly, the material will not bridge the opening and the watertight integrity of the surface coat is damaged. Such a coating may provide a good underlayment for another more flexible material, but in most cases, paint-ons should be viewed with caution if they are proposed as the only waterproofing material. If you have a rigid concrete shell

Photo 6-1: This self-adhesive waterproofing membrane is made of polyethylene with rubberized asphalt on one side. A protection board is applied to the membrane prior to backfilling or the addition of construction materials.

that is built to all but eliminate cracks, and the site is well drained, I would not rule out these paint-on materials. Follow the manufacturer's instructions carefully, and be sure to apply the materials only to clean, dry surfaces. Some of the paint-on materials are very flammable and have toxic fumes, so be careful when using them. Also, some of the coatings should not be applied to concrete that has been sprayed with oil-based curing compounds. There are some new paint-on materials that are supposed to be able to bridge cracks of up to $\frac{1}{16}$ inch without tearing. They are applied in a thick coat and when cured provide protection equivalent to a seamless membrane — unless a crack over $\frac{1}{16}$ inch develops. There are also paint-on coatings that appear to have as much stretch as membrane materials — enough to bridge a $\frac{1}{4}$-inch crack. The fly in the ointment comes from the fact that the paint-on materials bond to the surface, so only a very limited area of the material can stretch, while the membrane can absorb the strain over a much larger distance, since it is not, or is only slightly, bonded to the surface. The key word is bonded. If the directions for the material you are considering say that it *bonds* to the structural surface, then the material will probably split if a crack occurs. Since the concrete is essentially waterproof by itself, except where cracks appear, then a new crack will be the worst place for the waterproofing to fail.

Preformed Membranes

Membrane waterproofing is currently very popular for underground construction. Perhaps the least expensive of the membrane materials are the *modified bitumens*. They are essentially a combination of synthetic rubber and tar. They are generally sold in rolls like roofing felt, and they are more easily applied by the owner-builder than the large elastomer sheets. There are many seams, however, which must be sealed by lapping and gluing with a compatible mastic. Some products have an adhesive present in the material that seals to itself when properly lapped. Because the basic waterproofing material is not particularly strong in tension, a plastic film backing is often included to reduce the probability of damage while handling. The materials should be installed only on clean, dry surfaces and at temperatures usually above 40°F. Also, the material may degrade if exposed to sunlight, so you should backfill immediately after application. Multiple layers should be applied at corners, and although the material is somewhat self-sealing and will bridge cracks larger than those contained by paint-on materials, any area that has cracks already or is at the junction of slabs should also receive multiple layers.

Continuous membranes of butyl rubber (isobutylene isoprene) and a material called EPDM (ethylene propylene diene monomer) are available in widths of as much as 50 feet. Handling such large sheets without damaging them is not easy, so most owner-builders apply these materials over smaller areas with more seams rather than as a continuous sheet. Although there is no requirement (continued on page 160)

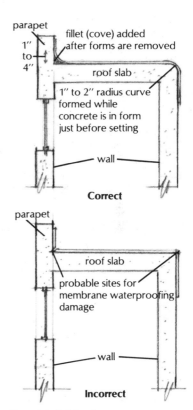

Figure 6–2: Membrane-type waterproofing is often damaged at sharp bends. The rounding of outside corners and the creation of a fillet (cove) at inside corners lessens the likelihood of damage.

Table 6-1
Waterproofing Materials

MATERIAL	COMMENTS	METHOD OF APPLICATION
Cementitious Products	Material is portland-cement based	Applied by spraying, brushing or troweling onto concrete
Liquid-Applied Membranes	Types include: acrylics; chlorosulfonated polyethylene (Hypalon); ethylene propylene diene monomer (EPDM); isobutylene isoprene (butyl rubber); neoprene (synthetic rubber); polyurethane; polyvinyl chloride (PVC); rubberized asphalt	Applied by trowel, squeegee, roller, brush or spray; dries to form a membrane that is continuously bonded to the surface
Montmorillonite Clay (bentonite)	Types include bulk bentonite clay, commonly called driller's mud and pond sealer; cartridge-mastic form of bentonite (Waterstop-Plus); panel form of bentonite (Volclay); spray-on form of bentonite (Bentonize R-80-S); trowel-on form of bentonite (Bentonize R-80-T)	Applied as a powder, spray, mastic or preformed sheets
Preformed Sheet Membranes	Types include: acrylics; chlorosulfonated polyethylene (Hypalon); ethylene propylene diene monomer (EPDM); isobutylene isoprene (butyl rubber); neoprene (synthetic rubber); polyvinyl chloride (PVC)	Provided as roll or cut sheet stock, which may be installed with no, partial or full adherence to the surface to be waterproofed

ADVANTAGES	DISADVANTAGES	EXAMPLES (TRADE NAMES)*
Less fragile than most waterproofing materials; easy to apply; relatively inexpensive	Useless if cracks appear after application, since it is not self-healing; since it's rigid, it can't bridge even tiny cracks; concrete surface must be clean before application and temperature must be within a relatively narrow range	Anchor Masonry Surfacer Foundation Coat Hey'Di K-11 Thoroseal
Result is a seamless, relatively easy-to-install membrane that easily seals concrete joints and cracks and around protrusions	Usually will split if the concrete cracks or shifts more than a fraction of an inch after the membrane has been placed, because it is continuously bonded to the surface; some contain toxic or flammable solvents and must be applied in well-ventilated spaces away from open flame; membrane works best if it is neither too thin nor too thick, and controlling thickness of application requires care; can be applied only if surface is clean and dry and if temperature is neither too warm nor too cold for the material being used; usually not self-healing, so even the tiniest tear or puncture will be a permanent flaw	A H Seamless Membrane Duraflex M Duralastic Duramen Extra Tite Gacoflex Geocel Exterior Brushable Sealant Geocel Water Shield HLM 1,000 Series HLM 3,000 Series Hypalon Liquiseal One-Kote Raylite Tremproof 50 Tremproof 60 Vulkem Wall Up
Is self-healing because it swells dramatically when in contact with water; may be installed over a wide range of temperatures; may be placed in greater thicknesses for problem areas; has no seams; nontoxic; increases its water resistance with an increase in water pressure	Will wash away in the presence of running water, so it can't be exposed to rain or surface runoff, or to rapidly percolating water that might occur through adjacent gravel; performance is reduced in salty soil; easily damaged mechanically; should be protected during backfill by rigid insulation or a 6-mil (or thicker) polyethylene film	Bentonize R-80-S Bentonize R-80-T Volclay Waterstop-Plus
If applied with less than full adherence, most preformed membranes will easily bridge new cracks or shifted joints in concrete; resistant to mechanical damage; can be installed over a wide temperature range; some can be purchased in large enough single sheets to virtually eliminate the need for seams	Most aren't self-healing; more than one person required for installation; hard to seal around irregular surfaces; easily damaged at corners; usually require seaming; some difficult to use on vertical surfaces	Hypalon Sure-Seal Vaporseal

(continued on next page)

*See Appendix 2 for addresses and phone numbers of manufacturers of waterproofing products.

Table 6-1—*Continued*

MATERIAL	COMMENTS	METHOD OF APPLICATION
Roll Goods	Purchased in rolls; made from several materials bonded together, thereby offsetting disadvantages of each; types include composite sheet membranes (built-up membranes), modified bitumens, rubberized asphalt	Generally installed fully adherent to the surface to be waterproofed

that the sheets be bonded to the concrete, if water gets behind a nonbonded sheet, it can travel long distances under the membrane before penetrating the concrete at a joint or crack. Obviously, finding the actual failure point in the waterproofing membrane is going to be pure luck in such a case. On the other hand, if the sheet is bonded continuously to the concrete, the result is the same as with paint-on materials: If a crack occurs, the material may tear. To control the flow of water under the membrane and still allow maximum stretch, the material may be bonded to the concrete in a cross-hatch pattern. As with the modified bitumen sheets, extra layers should be placed at stress points such as cold joints, corners and around roof penetrations. Where seams are necessary, special sealing materials provided by the membrane manufacturer are required. Waterproofing membranes forced into sharp inside corners such as at the base of a parapet are likely to split. It is good practice to add a concrete fillet to allow the membrane to make the corner gradually (see figure 6-2). Outside corners are less of a problem but where practical, a gradual bend is preferred. Beware of the power of solar energy when installing waterproofing membranes. Since the materials are usually black, they will heat up and stretch when exposed to the bright sun. If the membrane is installed when temperatures are above 60°F or on sunny days with temperatures above about 40°F, be sure to leave slack in seams and between strips where membranes are bonded to the surface. If sufficient slack is not present and if the membranes contract upon cooling, they may split. As Murphy's Law would have it, a membrane tightly installed on a hot day will be fine until covered by cool earth. Then the failure will be nicely hidden from view.

Montmorillonite Clays

Montmorillonite clays represent the third category of waterproofing materials. The common name for such clay is *bentonite.* While all clays tend to swell

ADVANTAGES	DISADVANTAGES	EXAMPLES (TRADE NAMES)*
Somewhat self-healing; can be applied over a wide temperature range; usually can be applied by single person with some difficulty and two persons easily; usually fully adherent, so water cannot travel behind membrane	Irregular protrusions harder to seal with composites than with bentonite or liquid waterproofing; membrane is fragile at sharp corners; seams may require special treatment; roll widths usually too small to allow placement without numerous seams; material is relatively heavy; installation on vertical surfaces more difficult than on horizontal areas	Bituthene Laurenco Melnar Poly-Mat Premolded Membrane Yellow Jacket

*See Appendix 2 for addresses and phone numbers of manufacturers of waterproofing products.

with increasing moisture content, bentonite does so quite spectacularly, with increases in volume of as much as 15 times its dry volume. As a result, when bentonite is spread over a surface, the presence of water causes it to swell against the surrounding earth, increasing its density and actually increasing its waterproofing capability. It is ideally suited to covering joints and cracks, since it swells to fill any exposed cavity. Bentonite can be applied in several ways. It can be troweled on, sprayed on or applied in the form of panels made of a layer of the clay with paper on both sides. When troweled or sprayed on, the clay powder is mixed with a special binder and *not with water*. I know of one builder who bought a large quantity of bentonite packaged for use as sealant for pond bottoms. He mixed it with water and applied it to the house shell. Of course, the bentonite was in its expanded form when it was put in place, so when it dried out, it shrunk 15 times, leaving rather large gaps over the entire surface. For the owner-builder, bentonite panels (trade named Volclay) and the troweled-on version called Bentonize are suitable. One special requirement of this material, however, is that it should be protected from contact with running water and from mechanical damage during backfilling. Covering the material with insulation or a polyethylene film where there is no insulation will do the trick. If you are building where the lower walls do not need insulation, you should either backfill with great care or place a more rigid protection than a poly film over the bentonite. Some builders use corrugated cardboard, and one I know uses ⅛-inch masonite, but my preference is to use ½ inch of foam insulation. The modest thermal isolation it provides is useful where ground temperatures can fall below 60°F, and it is good, economical protection for the waterproofing.

Waterproofing Block Structures

In general, concrete blocks are much less waterproof than poured concrete, so while a poured slab will leak only at cracks and cold joints, seepage through

Photo 6-2: A waterproofing membrane of EPDM is being applied to the poured concrete roof of this underground house. Two layers of 2-inch extruded polystyrene foam insulation were then applied.

blocks and mortar joints will be a problem over the whole wall if it isn't adequately protected. Some types of block will even wick water by capillary action through to the inner surface. While the same waterproofing methods that are used for any concrete wall will be appropriate for block walls, careful application and special care to prevent damage during the backfill operation will help eliminate problems of moisture entry through the blocks. The fiberglass-reinforced epoxy coating used to reinforce dry-stacked blocks is waterproof, but regular waterproofing treatment should be applied as well, since the joints between the floor, wall and roof are not waterproof. The coating material is designed as a structural material, not as waterproofing.

Waterproofing the AWWF House

As with all building systems, waterproofing works best when adequate drainage keeps standing water away from the house. The usual waterproofing method for wood basements is to caulk all seams and cover the outside of the treated plywood sheathing with a continuous layer of 6-mil polyethylene. Likewise, the floor is laid over a continuous sheet of poly. For the underground house, a less fragile membrane system or bentonite clay would be a better choice than 6-mil poly. The underfloor and wall surface drainage provided by gravel layers is very important, since it is very difficult to seal the junction of the bottom of the wall and the floor against standing-water leakage. In climates where the ground is likely to become saturated at some time or another, an additional line of drains installed at the junction of the roof and the wall (another place that is hard to seal) may be of value. In applying waterproofing, the same approach is used as

Photo 6–3 (left): *Bentonite waterproofing materials can be applied in powder, spray, mastic or panel form. Here bentonite waterproofing material is being sprayed onto a concrete surface.*

Photo 6–4 (above): *Bentonite waterproofing can be applied in panel form.*

for concrete, but don't forget that if you use interior insulation in the wood wall, the waterproofing will not be physically protected by a layer of insulation, and backfilling must be done very carefully. Remember, also, that if bentonite is used, running water should never be present at the wall surface. Covering the waterproofing with a sheet of polyethylene will both help protect it from damage during backfill and keep moving water from its surface.

Because wood is not as dimensionally stable as concrete, waterproofing is more difficult, and whatever the waterproofing substance, it must be capable of bridging gaps of ¼ inch and more, either by stretching or by self-sealing. Paint-on waterproofing will not work, since it cannot bridge large cracks.

Summary of Waterproofing Methods

The best waterproofing is drainage both at the surface and at the footing depth. Surface contours should be arranged to keep runoff flows channeled at least 10 feet away from the house, and no standing water should be allowed uphill of the house for 50 feet or more. There should not be standing water over the house site. Subsurface drains should be at the

(continued on next page)

Summary of Waterproofing Methods—*Continued*
bottom of a gravel or sandy soil fill that extends to at least the level of the roof in a completely underground house or within a foot of the surface in houses that are partially above grade. Drainage tiles or perforated pipes should be bedded in gravel larger than the openings of the pipes and must slope toward an open-air exit. Drainage tiles should be placed at least around the perimeter at or below footing depth. Sites that have all four walls below grade and floor spans greater than about 20 feet should probably have drains under the floor.

The material used to waterproof the shell should have the capability of bridging any cracks that may occur. All waterproofing materials should be installed during cool weather, or else there should be sufficient slack to allow for shrinkage as cooling occurs. A grid of adhesive should bond continuous membranes to the surface to isolate water that penetrates the membrane, while allowing flexibility. Corners, penetrations and cold joints in the concrete require special attention and there should be multiple layers of membrane materials or an extra thickness of bentonite. All waterproofing materials should be protected from mechanical damage by workmen, machines and the placement of backfill. Membrane materials are particularly likely to fail at 90-degree inside bends, so provide a fillet of concrete to break up inside corners where membrane waterproofing is applied.

Insulation

"Do you mean to tell me that I have to insulate my underground home? The earth is excellent insulation. Why insulate?" I hear comments like these at most underground house design seminars. The questions are valid. Not all sites need the same insulation. Not all designers or builders recommend the same amounts—some even recommend none at all. Indeed, if your underground house is in a climate where the winters are mild and cooling is a major problem, insulation requirements are much different from those of Saskatchewan. In most cases, though, leaving out insulation will create more problems than can be justified by saving a few dollars.

It is incorrect to consider earth as a good insulator. What the earth does is help *isolate* underground houses from the extremes of surface weather. This is due to the ability of a large mass of earth to store heat. The storage of heat in a mass causes a reduction in temperature swings because the mass can absorb great amounts of heat before its temperature rises significantly. Likewise, the mass can give off a lot of heat before its temperature falls very much. Since mass acts thermally as a flywheel acts mechanically, it is often called a *thermal flywheel*. Thermal flywheeling causes a leveling effect on temperatures by averaging the surface-temperature highs and lows. The flywheel effect also causes a delay between the time when low temperatures occur on the surface and when they appear at depth. The deeper into the earth, the more both phenomena affect temperatures. In fact, it is possible to reach a depth where the delay becomes a year or more, and the averaging effect yields a temperature that is the average for the entire year. This depth is called the *constant-temperature depth,* since the temperature is unchanging. Because this constant temperature is primarily an average of the year's surface temperatures, its value varies with latitude (and with elevation in mountainous regions). Figure 4–9 gives some indication of temperatures at various constant-temperature depths.

Of course, what you really have is a climate underground that varies less than the surface climate and generally much more slowly, but that climate is usually not within the comfort range, even at constant-temperature depths, for the majority of North America. Also, the picture is not complete until we know at what depth constant-temperature conditions occur and how the soil temperature varies above that depth. Quite a lot of temperature variation occurs in the top 3 feet of earth. Below that, the variation becomes less and less, until at about 10 feet, reasonably constant temperature conditions occur. The actual depth for constant temperature varies as a function of the temperature difference between seasons. In coastal areas, there is much less difference between summer and winter temperatures than in most mid-continent locations, so constant-temperature depths in coastal Oregon tend to be significantly shallower than on the Oregon-Idaho border. Since the subsurface temperature environment is the true "climate" around an underground house, understanding it is important to making an accurate assessment of insulation needs.

Let's assume that you do not include any insulation around your underground house. What are the potential problems? Since temperatures near the surface are not as high as those deeper in the ground, ceilings and upper walls will tend to lose an excess of heat in the winter, possibly causing condensation problems, higher utility bills and the generally chilly and damp feelings often attributed to basements and some underground dwellings. In the summer, those same surfaces will probably be warmer than is appropriate for comfort. Most underground-house designs include at least one wall that is exposed to the weather. Where the walls, roof and floor meet that exposed surface, the temperature of the adjacent earth varies along with outdoor temperatures. Even if the air temperature is in the comfort range, a room with cold walls will feel chilly. The reason for the chilly feeling is that the wall surface becomes cool enough to absorb heat radiated from the bodies of the occupants. Even worse, rooms with high-humidity conditions, such as bathrooms and kitchens, can have condensation problems due to wall temperature below the dew point of the room's air. For example, in a room at 70°F and with walls at 60°F, condensation will occur if the relative humidity rises above 70 percent. For most underground houses, there is justification for using some insulation. The questions that must be

Figure 6–3: The amount of insulation (in R-values) typically required for an underground house in the southern United States is R-16 within 2 to 4 feet of the surface (a quarter of the way down the wall) and R-8 within 4 to 6 feet of the surface (halfway down the wall).

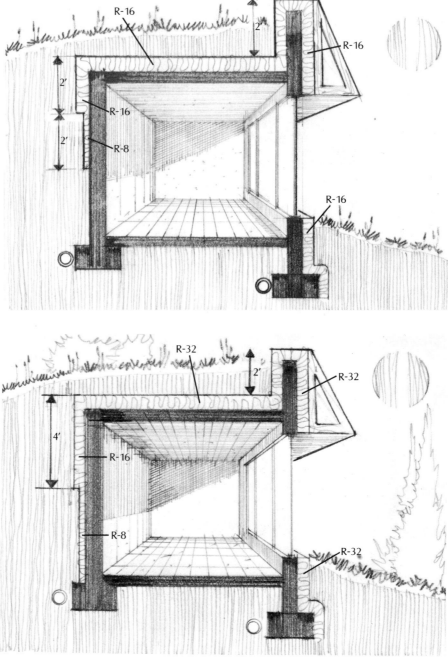

Figure 6–4: The amount of insulation (in R-values) typically required for an underground house in the central United States is R-32 within 2 feet of the surface, R-16 within 2 to 6 feet of the surface and R-8 over other areas.

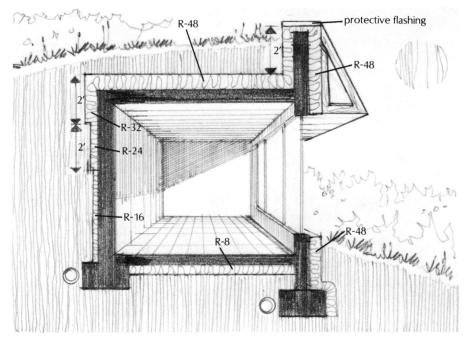

Labels on figure: R-48, protective flashing, 2', R-48, 2', R-32, 2', R-24, R-16, R-8, R-48

*Figure 6-5: The amount of insula-
tion (in R-values) typically required
for an underground house in
southern Canada is R-48 within 2
feet of the surface, R-32 within 2
to 4 feet, R-24 within 4 to 6 feet,
R-16 beyond 6 feet from the surface to the footings and R-8 under
the floor.*

answered for each site are how much, what kind to use and where to put it.

Since the temperature environment changes with depth, so should the amount of insulation. Those parts of the building skin close to the surface benefit the most from insulation. More insulation is usually placed on the roof and over exposed walls than around the lower parts of deeply buried walls. For subterranean climates where the constant-temperature level is below 65°F, some insulation is required all the way to the footings. For areas where the constant temperature level is above 65°F, there is justification for not insulating the deeper parts of the house. Almost everywhere, insulation is needed for the roof and any surfaces that are closer than 3 feet from the outside environment. It is generally not economical to insulate under floors if the earth temperature is consistently above 50°F but insulation under the floor is appropriate near outside walls, and there should be insulation at least 3 feet from any exposed surface.

Figures 6-3, 6-4 and 6-5 show the same house in three locations. The house in figure 6-3 is in the southern United States, the house in figure 6-4 is in the central United States, and the house in figure 6-5 is in southern Canada. Notice that each has the greatest insulation nearest the surface. Only the Canadian house includes any under-floor insulation, and that may not be really necessary. In each case, the designer wants to keep the interior surfaces of the houses at comfortable temperatures without losing too much of the summer cooling benefits that come from the cool earth. In the southern house, the earth is at 65 to 70°F at constant-temperature depth, so only the part of the house near the surface is insulated. Below about 4 feet, the walls and floor are in contact with the mass of the earth, providing rapid heat transfer from the house to the earth in summer and from the earth to the house in winter. Because the earth temperature (in contact with the uninsulated part of the house) is probably never lower than 65°F, condensation should not be a problem if there are reasonable precautions to remove excessive water vapor by adequate ventilation.

The house in the central United States has 4 inches of extruded polystyrene on the roof, 2 inches about halfway down the walls and 1 inch to the floor level (the assumed R-value for extruded polystyrene is R-8 per inch). The Canadian model uses the same graduated approach but has 6 inches on the roof, 4 inches on the top 2 feet of wall, 3 inches for the next 2 feet and 2 inches the rest of the way to the footings. The 1 inch of insulation under the floor reduces heat loss through the floor, but if the floors have carpeting or wood flooring over the concrete, the insulation may not contribute very much to comfort.

Notice that in each of the three examples, there is extra insulation near the exposed wall — including insulation as far down as the exposed wall's footings. Failure to increase the insulation near exposed surfaces is common and can result in sweating at corners and cold spots in the rooms nearest the outside wall.

Exterior Insulation

In addition to determining the amount of insulation, it is necessary to decide whether to place it between the house skin and the earth or to place it inside the house, as is usually done in aboveground houses. Most designers agree that exterior insulation is best for concrete construction. Having the insulation outside places a large amount of thermal mass (the concrete roof, walls and floor) inside the thermal envelope of the house. If the insulation were inside, that same thermal mass would be unavailable for a thermal flywheel effect. Exterior insulation also acts to protect the concrete's waterproofing material from damage during the backfilling operation. The primary disadvantages of exterior insulation are its cost and the difficulty of keeping it in place before and during the backfilling. Keeping the insulation firmly in place, particularly when upper parts of the wall have thicker insulation than the lower, is often a problem.

There are several possible ways to keep insulation in place. In one method, form ties are left in place instead of being broken off. The rigid insulation is pressed onto the protruding ties to hold it in place. Unfortunately, with the ties protruding from the concrete, some types of waterproofing — mainly membrane types — are both harder to install and more likely to fail. The continuous rusting of the ties makes long-term waterproofing almost impossible, and this method is not recommended. Some builders try to glue the panels in place, but that is usually not satisfactory. Common building adhesives that contain petroleum-based solvents can't be used, since petroleum products tend to dissolve polystyrene foam. Silicone rubber seems to be one of the few products that damages neither

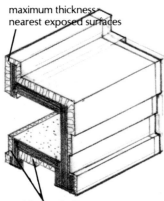

maximum thickness
nearest exposed surfaces

added underfloor
or footing insulation
where near exterior

Figure 6-6: Increased thicknesses of insulation must be placed on all surfaces near exposed areas, where most of an underground building's heat loss occurs.

the waterproofing nor the insulation and has some chance of sticking to both. Its cost is prohibitive as a general adhesive, however. Some foam manufacturers provide adhesives designed for polystyrene foam products. The Dow Chemical Company's Mastic #11 is such a material. Concrete nails spaced 16 to 24 inches apart can be used to fasten the foam to the concrete before backfilling. Nailing is recommended only for the person who can always hit nails squarely on the head. Just a few misses can seriously damage large areas of the insulation. A system used for insulating footings involves placing the insulation in the form when the concrete is poured. After the forms are removed, the insulation tends to stay with the concrete. This is not generally practical for surfaces that need to be waterproofed, since it is usually not practical to put waterproofing over the insulation. The exception is with some types of membrane waterproofing. Actually, the most common method of installation is to have someone hold each panel in place while enough fill is added to secure the bottom of the insulation. Then, as the backfill proceeds, a man using a T-brace keeps the insulation panels tight against the wall while the final backfilling is done (see figure 6-7). Pressing the panel against the wall during backfilling is important, since a common problem is getting rock and earth between the insulation and the wall. That causes damage to the insulation and possibly to the waterproofing as the fill forces the panels to conform to the uneven surface. Try to backfill immediately after placing foam insulation. Even slight winds may catch the poorly supported panels and tear them loose. You haven't seen activity until you have been at a homesite where the builder has fixed *only* the base of the insulation panels and a wind comes up! If multiple layers of insulation are installed, stagger the joints of each layer for best insulation performance. Roof insulation, in particular, needs to be protected from damage. While it is possible to walk on foam insulation without damaging it, this is not advisable. Covering the insulation with heavy cardboard, masonite or plywood where walkways are needed will help preserve that expensive R-value for your roof. Discourage people with wheelbarrows or other wheeled machines from working on unprotected surfaces, and be aware that dump trucks, bulldozers and backhoes can place rather deep dents in extruded polystyrene.

Figure 6-7: Although adhesives can be used to hold insulation against most surfaces, using a homemade T-brace is usually practical and allows concurrent placement of insulation and backfill.

Interior Insulation

If the house's skin is wood or metal, interior insulation is my first choice. While the main reason exterior insulation is preferred for concrete construction is that thermal mass is left next to the living space, low-mass construction does not benefit greatly from exterior insulation. Interior insulation is certainly easier to apply on the walls. Fiberglass or cellulose insulation costs less than rigid foam and the space provided for insulation can double as plumbing and wiring runs.

Using interior insulation on concrete may be cheaper and allows more latitude in interior finishes, but it removes the thermal mass from the room. Some builders compromise by putting interior insulation on some walls and exterior

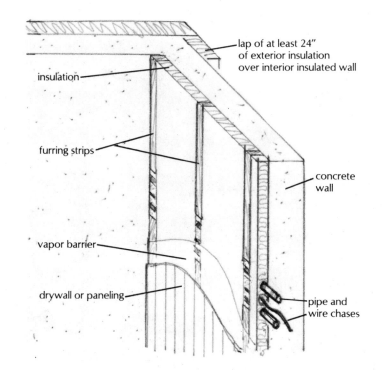

lap of at least 24"
of exterior insulation
over interior insulated wall

insulation

furring strips

concrete
wall

vapor barrier

drywall or paneling

pipe and
wire chases

*Figure 6–8: Walls not used for so-
lar heat storage are often furred
out and provided with interior insulation. This provides space for
plumbing and wiring without using
conduit. To prevent large heat
losses from occurring at the junction of surfaces with exterior insulation, such walls should have exterior insulation installed at least 24
inches beyond the junction.*

insulation on others. Any walls that are exposed to direct sunshine should have
exterior insulation both to store heat and to keep room temperatures from rising
too fast. On the other hand, north walls — particularly those in utility areas, closets
or other nonliving space — will be fine with inside insulation. So, a reasonable
compromise is to use interior insulation on the north wall of a passive solar
underground house and exterior insulation elsewhere. If that is done, be sure to
remember to continue the exterior insulation several feet onto the north wall so
that the corners are not left uninsulated (see figure 6–8).

Adding Mass with Insulation

There is a third place to put insulation; it is not popular, but in the right
situation it can be quite effective. That is neither on the inner nor the outer
surface of the house skin but rather parallel to it and a foot or so away in the earth
(see figure 6–9). That effectively adds thermal mass to the house's thermal
envelope. In a concrete house, that mass is not necessary, nor even particularly
useful in most cases, but for the wood- or steel-shelled house, it can provide the
same thermal flywheel effect as in a concrete house, and at considerably reduced
cost. Unfortunately, it is very difficult to place foam panels away from a
supporting surface, keep the edges together and do a good job of backfilling.
Frankly, I don't know an easy way to use this method of increasing thermal mass,

unless you happen to have a site where the soil can be sheared off vertically and kept that way. Then you may be able to put the insulation up against the earth and fill between the insulation and the house. The hard part is the backfilling. Most sites require a lot of slow backfilling, mostly done by hand. But if you want the most comfort for your metal- or wood-skinned home, that is one way to get it.

What Kind of Insulation Is Best?

A generally controversial subject is what kind of insulation to use for exterior insulation. Almost everyone agrees that polystyrene insulation is the best choice for insulation in contact with the earth. It is not biodegradable, and it is mechanically stable. But, should you use the low-cost white beadboard, which is made of polystyrene beads that have been expanded and then pressed together, or the more expensive extruded polystyrene, which is sold under the trade names Styrofoam (blueboard) and Foamular (pinkboard)? No one seriously argues against the extruded forms as far as thermal performance is concerned. The problem is cost. The extruded forms are structurally stronger and have a higher R-value per inch than beadboard, although beadboard has a lower cost per R. Why not beadboard then? The problem is water absorption and long-term mechanical damage. Manufacturers of extruded-polystyrene insulation claim that water absorption in beadboard causes as much as a 40 percent reduction in insulation capability. Beadboard manufacturers say that statement is true only if the material is soaked in standing water for long periods, and that will never happen in a real installation. To further complicate matters, the water absorption of beadboard seems to vary considerably among manufacturers. Then, too, for some sites, the constant swelling and shrinking of the surrounding soil with changing moisture content supposedly gradually separates the beads and destroys the insulation's effectiveness. Since the time scale for this to happen is measured in years, no really adequate experimental data is available for different soils in different climates. Until unbiased, adequate test data become available, the general attitude of those in the building industry is to wait and see.

A few builders are placing beadboard next to the concrete and using extruded polystyrene only on the outside layer, which is in contact with the earth. A few others are using all beadboard and protecting it from the earth with various materials ranging from building paper to plywood. All have worked over the short haul. How they last over the long haul is still unknown. Certainly, the safest approach continues to be to use only extruded polystyrene as external insulation.

Insulation for Ferrocement Construction

When ferrocement construction includes curves, adding insulation becomes a challenge. Insulation placed outside the concrete must be a closed-cell, non-water-absorbing material that will not degrade underground. Only

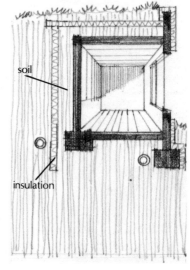

Figure 6–9: Insulation can be installed a foot or so away from and parallel to a wall to increase thermal storage, but this is a technique that is easier to illustrate than to do!

extruded-polystyrene foam fits this description. I know of no foam-in-place materials that meet all those needs, so laying up multiple layers of board-type extruded-foam insulation seems to be the best way to get a complex surface properly insulated on the outside. Foam boards will bend enough to follow room-sized curves on domes and walls. If these ¾-inch or 1-inch panels are put on in layers so that the joints are staggered, the end result should be comparable to a single thick layer. The only placement method I know that works well is to start at the bottom of the wall or dome and backfill as you place all layers. Otherwise you will find it almost impossible to keep all the panels in place unless you use hundreds of dollars worth of adhesive. Even with the backfill-as-you-go method, several workers will be needed to hold the insulation in place while fill is added.

Domes with nonstructural interior finishes can have foam-in-place insulation applied between the interior dome shell and the ferrocement armature. After curing, the foam insulation should be covered with plastic before the concrete application, however, so it will not absorb moisture from the concrete. Some houses have been built that have foam insulation sprayed on the outside of the dome. The insulation is covered with sheet plastic to isolate it from the soil. Spray-on foam insulation (urethane only) can also be added inside the dome surface, but I would consider that to be a fire hazard unless it in turn is covered with a layer of plaster or some fire-retardant material.

Insulation Design

Find the depth at which constant temperatures occur at your site. Check with a university geology department or the state agency concerned with geology or mines. If you can't get that data for your area, find out what the average temperatures are for the warmest and the coldest months. If they differ by more than 50°F, 10 feet is a reasonable guess. If they differ by 30 to 50°F, 5 to 10 feet is about right. If they differ by 10 to 30°F, 2 to 5 feet is approximate. If a difference of less than 10°F occurs, then assume 2 feet.

Determine the constant-temperature value. The map in figure 4–9 will help you unless your house is in the mountains, though again, universities and state agencies may be of some help. The temperature of well water is usually accurate, also.

Once this temperature and depth information is available, intelligent placement of the proper amount of insulation becomes easier. Although the actual temperature distribution from the constant-temperature depth to the surface is a complex, dynamic function, a conservative method of figuring the required insulation is to use the model house diagrams (figures 6–3, 6–4 and 6–5) and place the thickest insulation at those parts of the structure that are closer to the surface than halfway to the constant-temperature depth. For the rest of the way to the constant-temperature depth, cut the thickness in half. Below the constant-temperature depth, use no insulation if the temperature is 60 to 75°F, and use R-4 to R-8 insulation for temperatures from 50 to 60°F. For temperatures between 40 and 50°F, R-8 to R-16 insulation is adequate. For houses in very cold climates, where the temperature at constant depth is below 40°F, R-24 to R-36 insulation provides the best insulation value.

Summary of Insulation

Decide where to place exterior and interior insulation. Any concrete that has its inside wall exposed to the sun will provide great benefit if exterior insulation is used. Concrete roof structures are usually cheaper and perform better thermally if insulated on the exterior. Exterior walls that are not in sunny areas (particularly if they are in nonliving areas of the house) may be insulated on the inside with little penalty thermally and usually at lower cost. Walls that include many plumbing and wiring runs may be best furred out and insulated on the inside. Wooden or metal skins are usually easiest to insulate on the inside, but considerable thermal storage benefit sometimes can be obtained by placing rigid insulation in the earth a foot or so away from the exterior of these lightweight walls.

Concrete structures are more comfortable and perform better thermally if exterior insulation is used. If your house is concrete and the constant-temperature regime is 60°F or higher, the insulation plan in figure 6–3 is satisfactory. If your constant-temperature regime is 50 to 60°F, the insulation plan in figure 6–4 is appropriate. If constant-temperature levels are below 50°F, the plan in figure 6–5 is optimal.

The minimum thickness of exterior insulation is needed below the constant-temperature depth, with increasing thicknesses near the surface. The actual thickness depends on the underground "climate," but in all cases, maximum thicknesses should be used closest to the surface.

The preferred material for exterior insulation is extruded polystyrene. Beadboard *may* work fine, but then again, it may not.

For interior insulation, the R-values should be no less than those recommended for exterior insulation. Since it is usually not practical to change the R-value of interior insulation over a given surface, the whole surface should be insulated to the level needed for the worst case. That is, on a wall, the insulation should be thick enough to take care of that part of the wall nearest the surface. Methods for installing interior insulation are the same as for aboveground houses.

CHAPTER 7

Subterranean Subsystems: Wiring, Plumbing, Heating, Cooling, Ventilation and Lighting

In designing a space station, NASA engineers spend a lot of time on life-support systems — those systems that ensure the overall health of the astronauts. The total life-support system has several subsystems — air, light, food, water, waste disposal and thermal control — each of which is carefully integrated with the others. The same life-support systems are just as important for people in houses as they are for people in space. Yet, the selection of the subsystems for domestic life support is usually arbitrary and done with little regard for any interactions that may occur. Typically, a lot more thought goes into the selection of the color of the paint in the dining room than into the selection of a furnace, a ventilating system or a kitchen range. Yet, the effect on the occupants of a poorly chosen furnace or ventilation system is greater and much more difficult to correct than the color of paint. It is a shame to build an underground or any type of house that has the potential for a high level of comfort and energy efficiency and then introduce unnecessary drafts, noises, temperature fluctuations, odors and other unpleasantness because of poor choices in subsystems. However, a little planning and forethought are all that are needed to assemble the optimum subsystems that will allow the best possible house performance.

Electrical Systems

For underground housing, the electrical system must be carefully considered prior to construction of the shell, especially if the shell is to be concrete. In a standard wood-frame house, wiring is simply run willy-nilly through studs, rafters and joists. The house framer and the electrician probably never see each other, let alone discuss the most effective way for their respective responsibilities to interact. Unfortunately, many underground houses are built the same way. The shell is constructed out of concrete, forms are stripped, interior walls are framed, and then the electrician is called and told to "wire it up." With no stud walls, no rafters and no floor joists, there is a problem. The usual resolution is to build a wooden wall a few inches thick just inside the concrete walls, run the wires there and cover them up with paneling or drywall. Likewise, a suspended ceiling may be the result of poor planning for overhead wiring needs. Yet, there is little economy in building a wall inside of another wall, particularly when the wooden wall effectively reduces the comfort of the home by isolating the temperature-moderating concrete from the living space. Adding a suspended ceiling under an 8-foot concrete ceiling creates unpleasantly low ceiling height, with associated claustrophobic effects. Raising the concrete roof so the ceiling can be lowered to 8 feet by

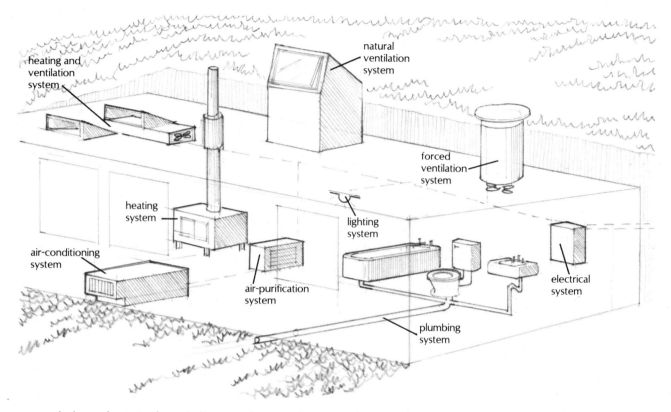

Labels in figure:
- heating and ventilation system
- natural ventilation system
- forced ventilation system
- heating system
- lighting system
- air-conditioning system
- air-purification system
- electrical system
- plumbing system

Figure 7-1: Comfort and convenience are dependent upon the interactions of many specialized home subsystems.

suspended panels costs plenty in increased excavation costs, increased concrete costs and the cost of the extra ceiling itself. All those problems are avoidable; what is required is planning. During the planning stage for the house, considerable money can be saved if the electrical system is completely defined along with the rest of the house.

By using conduit (usually steel; sometimes plastic), wiring paths are formed within the concrete. This eliminates the need for having furred-out walls to run wiring and allows the concrete mass to remain in thermal contact with the room — a significant advantage. But TANSTAAFL (There Ain't No Such Thing As A Free Lunch) persists, and placing conduit in this manner requires careful assembly, and it must be mechanically sound when the pour is made. You can't poke wires through concrete-filled conduit. It also absolutely requires careful teamwork between the electrician and the concrete workers. While I would not recommend a cast-in-concrete conduit system to the average do-it-yourselfer, it is useful to know what the potential problems are so that you can at least inspect the work to see how well it is being done.

Probably the most common installation problem is inadequate mounting of the conduit prior to the pour. When that concrete comes pouring down the

Photo 7-1: Electrical conduits formed in concrete can eliminate the need for furring out walls to accommodate electrical wires or for using surface wiring systems, but care must be taken to install the conduits properly so that they aren't dislodged or damaged when the concrete is poured. Electrical conduits are installed in this tilt-up wall.

form, a lot of pressure and stress are placed on anything in its way. Wiring the conduit to the concrete reinforcing rods every few feet is good practice. In particular, outlet boxes must be secured, preferably to the forms. Even if the outlet box does not separate from the connecting conduit during the pour, if it is significantly shifted, it may end up at a different height from the others along the same wall or be canted or otherwise cause a poor-looking installation. The box may even move deeper into the concrete mass, allowing it to be covered and making it particularly difficult to reach. Then, if it can be accessed, installing outlets will be a problem. Joints are the weakest part of the system and should always be fastened to the rebar or the forms. The joints must be made with what are called concrete-tight fittings. If it sounds like a lot of trouble, it can be, but in the long run it's the best way to do it in a concrete shell. Competent, experienced installers are the best insurance.

Casting conduit into the ceiling requires the same techniques used for the walls. Usually it is easier because the concrete is less likely to disturb the conduit or the outlet boxes as it is poured over a shallow surface than as it falls 7 or more feet down a deep wall form.

If you do decide to do the conduit work yourself, there are several important rules to follow. First of all, keep the bends few in number, and where they do occur, make them of large radius. Conduit bending tools are available and will make the appropriate wide bends. Second, use the correct hardware. If the wiring is to be placed in concrete, use concrete-tight connectors and boxes. Whether the conduit is cut with a special tool for the purpose or with a hacksaw, be sure to ream out the inside of the cut to remove any sharp burrs that might damage the insulation on the wires as they are drawn through. Finally, read the parts of the National Electrical Code that give information concerning wiring in conduit, such as the number of wires of various sizes that can be safely placed in each size of tubing and the types of insulation needed on the wires. By the way, don't expect to save much money by doing it yourself. I have seen cases where a do-it-yourselfer has spent more on materials alone than the total cost of a professional wiring job. On the other hand, electrical wiring is not particularly arcane, and there are many books on the subject (although few on conduit wiring) to guide you through the more confusing areas. One final admonition: In your planning, don't forget those electrical odds and ends like doorbells, intercoms and small fans to help passive solar units. After the concrete has set, 1 wire is just as painful to install as 12.

Probably the best way to eliminate problems with cast-in-place conduit is to use it sparingly. That is where careful design is important. Run as many concealed wires as possible outside the concrete. Install wiring in nonconcrete interior walls, behind built-in cabinets, in closets, in air duct chases and in ceiling beams, or use surface-mounted units wherever possible.

Surface-mounted wiring devices are not new, but they have been little used in houses. One system that is unobtrusive yet quite adequate for home installation places the outlets and the wires within a baseboard assembly. These look much like ordinary baseboard materials but are actually modules that include sockets for regular 110-volt plugs, corner connectors and blank extension sections. Other surface-mounted outlet strips range in appearance from sleek and handsome to conduit and box installations (for that high-tech look). A word of caution: Some of the "sleeker and more handsome" units have limited power-carrying capacity, so if they are to act as a passageway for power to

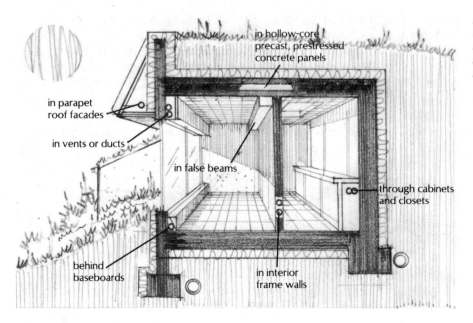

in hollow-core precast, prestressed concrete panels

in parapet roof facades

in vents or ducts

in false beams

through cabinets and closets

behind baseboards

in interior frame walls

Figure 7–2: With careful planning, electrical wiring can be installed in a concrete house without the need for much conduit. Here are some options.

Photo 7-2: Surface wiring systems, frequently used to add more electrical outlets to a kitchen, can be used in underground houses to avoid the need for cast-in-place conduits. These systems are attractive and are designed to meet a variety of electrical needs.

another room, be sure that enough current-carrying capacity is available. If it is practical to do so, use these low-capacity strips at the end of a circuit rather than in the beginning or middle. One firm that manufactures a variety of surface wiring systems is the Wiremold Company in West Hartford, Connecticut.

One system design method for underground houses places all the main feeder wires in a wooden chase disguised as a beam running down the long axis of the house. All ceiling-mounted fixtures such as lights and Casablanca fans are attached to the false beam rather than to the concrete. Where the "beam" intersects the interior walls, branch circuits leave and go to the front and rear of the house where the baseboard outlet system is mounted to the concrete surfaces. The main distribution panel is in a closet or utility room, and the only conduit required to be placed in concrete is for pass-throughs to the outside lights, exterior outlets and the incoming power lines.

Another method of concealing wiring uses a wainscot and chair-rail wall finish. In this approach, the wall is first furred out to a height of about 30 inches. The wires and special, shallow outlet boxes are then installed. The exposed concrete above the furring is plastered out to be about flush with the furred part, and the furring is covered with paneling. Finally, a chair rail is fastened at the top of the wainscot to cover the joint between the furring and the plaster. This system allows most of the thermal mass to stay within the living space but provides full freedom in wiring runs and outlet placement. This approach requires thicker plaster (if you want the wainscot flush with the wall surface), but that should not be as expensive as furring the entire wall.

A more common approach to the wiring problem is to form a wiring chase into the inside of the concrete walls at outlet level. A 1 × 2 run with a 2 × 4 block at the site of the outlet boxes, all fixed to the inside of the inner form, provide the necessary channel in the concrete surface. The wires are then run, and the wall is covered with drywall or paneling. The outlet boxes may be concreted in place, affixed to blocks, attached to the concrete in the wall openings or attached to the covering material. When the covering material is paneling, the thermal-mass effect of the concrete is greatly reduced. If drywall is used over the concrete, the loss is not as great. Probably the best way to finish the surface is to cover the wiring chase with kraft paper, fasten metal lath over it and plaster over the whole chase, being careful to keep plaster out of the chase itself. Don't overlook one important aspect with this wiring method — the possible effect of the cast-in-place wiring chase on the strength of the concrete. Tell the structural engineer in advance of his design work that this approach to wiring will be used. He can take into consideration the additional stress concentrations that occur because of the presence of a discontinuity in the wall surface. The engineer may recommend that the form for the chase be made rounded rather than square-cornered to reduce the likelihood of cracking at corners. In any case, the chase need not be very deep to contain the wires, and the shallower it is, the less effect it will have on wall strength. In particular, wiring chases in ceiling panels must be carefully designed to avoid structural problems.

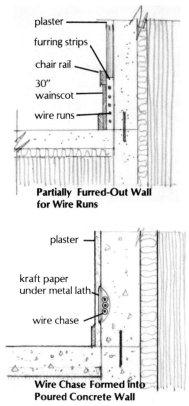

**Partially Furred-Out Wall
for Wire Runs**

**Wire Chase Formed into
Poured Concrete Wall**

Figure 7–3: Two methods for running electric wires in walls without conduit or without serious loss of thermal mass are to use a partially furred-out wall for wiring runs or to use a wiring chase formed into the poured-concrete wall.

Summary of Electrical Systems

Avoid conduit wiring as much as possible by carefully planning the wiring system to run in furred-out sections, through cabinets, through hollow beams, inside duct chases, under baseboards, inside surface wiring units, along parapet enclosures and in interior walls. Where wiring must be run through concrete, secure conduit adequately to the reinforcing bars and boxes to the form or reinforcing. Always use concrete-tight fittings and boxes. Have as few bends as possible in conduit runs, and where they are necessary, sweep them around a large radius. Observe limits recommended by the National Electrical Code for insulation type and number of wires in conduit. Consult your structural engineer to determine if formed-in-place wiring chases will modify wall or ceiling thicknesses. Above all, plan, plan, plan — and *use* the plan!

Plumbing

Critics may tell you that all a plumber needs to know is that sewage flows downhill. I say, t'ain't so. As with electrical systems, careful planning born of experience usually provides the best system at the least cost. Actually, under-

Figure 7–4: Plumbing pipes run under floor slabs should have all joints above the floor, and any pipes passing through the slab should be protected by a plastic sleeve. These precautions ensure that the expansion and contraction differences between the concrete and the pipe will not wear through a pipe or pull a joint apart.

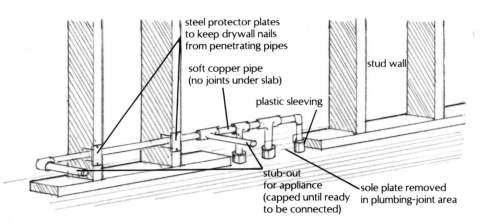

steel protector plates
to keep drywall nails
from penetrating pipes

soft copper pipe
(no joints under slab)

plastic sleeving

stud wall

stub-out
for appliance
(capped until ready
to be connected)

sole plate removed
in plumbing-joint area

ground houses differ very little from conventional houses that are built on a concrete slab, at least with regard to water supply. The system should be designed to keep plumbing out of the concrete, except in the floor, where any plumber with a reasonable amount of experience will know what to do. For your benefit, there are two main things to remember: Always put a plastic sleeve around pipes where they come through the slab, and never, never have pipe joints under the slab. Soft copper pipe is usually used to eliminate joints. It is buried in the earth or gravel under the slab, not laid on top of the earth where the concrete will pour around it. Wherever it is in contact with concrete, it is wrapped in plastic or inserted in a plastic sleeve. I have seen bendable plastic pipe (polybutylene) used under concrete floors, and I know of no problems, but I will admit to being leery of it. Under no circumstances should you allow galvanized pipe, rigid plastic pipe, or any other pipe that would require joints, to be placed under the slab. As figure 7–4 shows, where it is necessary to join pipes to run to other parts of the house, the ends of a continuous run of pipe are brought through the slab and the joint is made above the floor in a wall cavity or in the utility room, where the joint will not be in the way. Hot and cold water pipes are treated the same. Usually, there is little advantage to insulating hot water pipes unless the hot water is continuously circulated with a pump.

Water pressure is one thing you should always check. You only need about 25 pounds per square inch (psi). Pressure over 40 psi is asking for leaks and possible damage to clothes washers, dishwashers and other appliances that have electrically operated water valves. Remember that pressure in the water main is the *lowest* pressure in the system. Every time a faucet is turned off rapidly, a shock wave travels through the system that amplifies pressure of the water main several times. In one house I visited, the watermain pressure was 85 psi, and turning the tub faucet off too fast popped the overpressure valve on the water heater — dumping over 500 gallons of water onto the lawn. Measure the pressure of the water main. If it is over 40 pounds, consider installing a pressure-reducing valve on the inlet pipe. If it is over 75 pounds, run, don't walk, to the

Table 7-1 _____

Characteristics of Plastic Piping

TYPE OF PLASTIC	CHARACTERISTICS	COMMON USE	JOINT METHOD
ABS (acrylonitrile-buta-diene-styrene)	Rigid, with fair resistance to chemical action; withstands heat well up to 120°F; usually sold in 10-ft joints	Drainage and vent systems of all types	Solvent weld for permanent bonds; threaded adapters for removable joints
CPVC (chlorinated polyvinyl chloride)	Rigid, with good resistance to most chemicals; will work up to 180°F, higher than found in home DHW* systems; should not be fixed solidly in place due to its great dimension changes with temperature changes; sold in 10-ft joints	Hot-water supply for homes and general water-supply systems where higher pressure capability is needed (up to 400 psi)	Can be threaded but is usually solvent-welded
PB (polybutylene)	A very flexible pipe; now available in coils up to 100 ft long; various ratings for pressure are available	Water supply, both hot and cold; particularly suited to sites requiring lots of bends	Can be solvent-welded but is usually joined with clamp-type fittings
PE (polyethylene)	Flexible; sold in coils up to 500 ft in length; not for hot water but can be purchased to withstand pressures of 180 psi and more	Supply lines from wells or other underground water distribution networks like sprinklers and irrigation systems	Clamped joint fittings
PP (polypropylene)	Rigid and very resistant to chemical and mechanical damage	Fixture supply lines and drains; fixture traps; both hot and cold water	Threaded or compression-type
PVC (polyvinyl chloride)	Rigid, with good resistance to chemicals; withstands up to 110°F but may sag; great changes in length (up to ½ inch in 10 ft for a 100°F temperature change) and should not be cast in concrete or rigidly fixed in place; sold in 10-ft joints	Cold-water supply for homes and industry; used for gas piping in some underground installations; high-pressure pipe (up to 400 psi rating) is available; used for drains and vents	Can be threaded but is usually solvent-welded

*Domestic hot water

plumbing-supply store for one. Even if the pressure is below 40, lower water pressure usually results in reduced water use, and the pressure reducer may pay for itself anyway. My home is on a private water supply, and over the years I have settled on 25 psi as being a reasonable trade-off between too little and too much. At one time I had it down to 10 psi, which was fine for personal use, but the washer took a long time to fill.

Rigid plastic pipe has been used for years with success, and for the do-it-yourselfer, it is a real pleasure compared to working with copper or galvanized steel. It is, of course, more fragile and should be well protected against mechanical damage. Where it passes through studs, be sure to put steel protector plates over the stud if there is any chance at all that someone will drive a nail at that point. No matter what kind of piping system is used, be sure you pressure-test all parts of the system that will be hidden, before they are covered. If you place drywall over a pipe that leaks, Murphy or some kin of his must have a law that says that the water will always appear on the floor at least 10 feet away from the leak. The end result is usually the destruction of most of a wall that probably was the only perfect drywall job you've ever done. Although most plumbers crimp or cap the ends of rough-in pipes to keep debris out of the system, it is a good idea to flush out all lines before attaching fixtures. It is not at all unusual for rocks, solder, insects and other assorted doodads to be present in those supposedly spanking-clean pipes. They can easily cause problems with valves, faucet screens and other fittings. In one case, a neighbor child spent the better part of an afternoon listening to the neat sound marbles made as they rolled down a copper pipe.

The one area of plumbing that is potentially different from conventional housing is sewage disposal. For some homesites, the plumber has to make the sewage flow uphill! There are two cases where that can happen. The first is where the municipal public sewer system is not as deep as the house floor. The second, and less common, instance is where a private disposal system, a septic tank or similar system must be closer to the surface to keep leach fields out of impervious soil or to keep them above all groundwater conditions. In either case, special sewage-pumping systems must be installed, which is not a job for the do-it-yourself plumber. They can be tricky, and they are usually expensive and require attention to detail to work properly. Fortunately, when they are properly installed, they are usually maintenance-free, and the user can't perceive any difference between the "up-flush" and the "down-flush" systems. As far as the commonly voiced fear of a back-up from the main sewer is concerned, check valves (one-way valves) are adequate protection and are included as a part of the system.

What is the minimum fall needed in the sewer line before a pump is required? The rule of thumb is that the house sewer line should slope about ¼ inch per running foot. So, if the house is 100 feet from the main sewer line, the house sewer line has to fall at least 25 inches. While we are on this subject, popular books on plumbing do not always note the fact that waste and sewer lines can actually slope too much. What happens then is that the liquids run away so fast that the solids are left behind and cause clogging. If the main sewer is significantly deeper than the floor level of the home, the waste piping should be run at a slope of ¼ to ½ inch per foot to the location of the sewer line. Then it should be dropped steeply (45 degrees or greater) to the sewer.

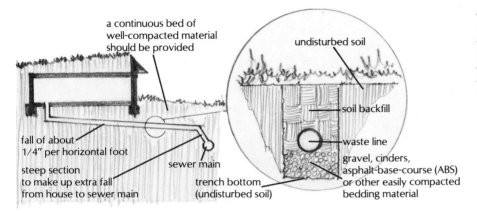

*Figure 7-5: Waste lines should be
well supported and properly
sloped. If a sewer main is too low
to connect with the house waste
line while maintaining the proper
slope, a short, steeply sloping section may be used at the main.*

Although most house waste lines work well at a ¼-inch slope per foot, the idea is to get wastes to flow at about 260 feet per minute — a velocity found to be just enough to keep solids moving but not so fast as to leave them high and dry. The following is a formula for grading drains:

$$F = \frac{L}{10d}$$

where F = total fall (grade) in feet

L = total length in feet

d = diameter of the pipe in inches

So a 4-inch drain, 100 feet long would have a fall of

$$F = \frac{100}{10 \times 4} = \frac{100}{40} = 2.5 \text{ feet}$$

That relates to a fall per foot of 2.5/100 = 0.025 feet of 0.3 inches per foot (very close to the rule of thumb of 0.25 inches per foot). What kind of pipe is best for the sewer line? Cast iron is the old standby, but nowadays plastic is probably the most used. Properly installed, the smooth plastic is less likely to clog than the rough cast iron. Joints are smoother, too, with less likelihood of catching solids. The main disadvantage of plastic is its relative fragility. It must be laid on a smooth bed, properly sloped, while cast iron will easily span dips in the trench. Care must be taken in backfilling plastic lines, or hidden damage may result. After it leaves the area of the house, vitreous clay tile is common for the sewer line, but cast iron and plastic may be used all the way to the sewer main (or lagoon, septic tank or

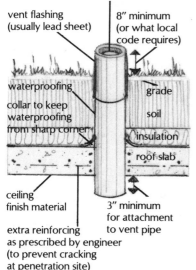

cast-iron vent pipe
(size depends on
number and type
of fixtures vented
and on building codes)

vent flashing
(usually lead sheet)

8" minimum
(or what local
code requires)

waterproofing

collar to keep
waterproofing
from sharp corner

grade

soil

insulation

roof slab

ceiling
finish material

3" minimum
for attachment
to vent pipe

extra reinforcing
as prescribed by engineer
(to prevent cracking
at penetration site)

Figure 7–6: Cast-iron vent pipes poured into the roof slab with proper reinforcing flashing and waterproofing will be leak-free and will require no maintenance. The pipe should extend into the room below just far enough to allow coupling with the vent system.

aerator box). It is good practice to bed sewer lines carefully, no matter what they are made of. That minimizes the likelihood of settling and resultant dips in the drain. By the way, if you ever do have a Roto-Rooter work over your drains, be sure to tell the operator if you have plastic pipe. If he treats your plastic drains like cast iron or vitreous clay, they could be damaged.

Vents are the last area of concern in plumbing. For proper action of drains, vents to the atmosphere are necessary. Building codes require such vents to exit outside the house. With the underground house, that usually means a penetration through the roof and a site for a possible leak. Use iron vent pipes where possible, because concrete will bond to the iron but not to plastic. Since the vent protrudes through your roof "lawn," plastic is also more likely to be damaged by lawn mowers and people. Obviously, the penetration needs to be made at the time the roof is cast in the case of poured concrete. What is usually done is to place a section of iron vent pipe that extends to its proper height aboveground but down into the interior space only 4 to 6 inches. Later on, the vent pipe is extended on to the attachment point at the plumbing "tree," where all the fixture drains come together. The reason for not running the vent all the way to the floor is that the plumbing tree is usually not built up until after the roof is poured — it might be damaged by the activity of the concrete workers placing and removing the forms and form shoring.

Where practical (and permitted by code), keep the number of vent penetrations to a minimum by tying together as many vents as you can. Just be sure that the vent ties are several feet above the level that entering water can reach from sinks, washers and other fixtures. Under no circumstances vent the plumbing to the interior of a wall or to an attic space. Sewage generates explosive and toxic methane gas (colorless and odorless), and this could easily leak from your vent into the house.

Composting, Waterless Toilets

The use of composting, or waterless, toilets in underground housing is fairly uncommon. Of course, in the housing stock as a whole they are also a rarity, but that shouldn't discourage you from considering them as a reasonable alternative, particularly if you are too low to drain into a sewer and a possibly expensive pumped sewage-disposal system is needed. Except in those rare two-story underground dwellings, the two-level composters are impractical for underground houses. The expense of excavating and enclosing a space below the already deep floor level is too great. However, some designs do not require a level below the bathroom. Some problems with insects and with venting have been reported, and before committing yourself to a particular design, try to locate and visit a family that has had one in use for awhile, preferably for several years. Remember that the low infiltration rates of underground housing can cause odors to linger longer and seem more pronounced than in conventional houses, so provide adequate ventilation. Some units have been in use for many years and are proven to be reliable, while others are new on the market. Choose carefully and you should not be disappointed.

Summary of Plumbing System

Try to choose a homesite that allows gravity entry to the sewage-disposal system. Pumped systems are available but costly and can be a problem if not installed correctly. In installing building sewer lines from the house to the sewer main and other disposal area, be sure the pipes are laid with a slope that is adequate for drainage but not so great as to encourage liquids to leave solids behind. Bed the sewer pipes carefully into well-compacted earth or gravel so that settling of sections of the pipe with resulting dips will be unlikely.

Use pressure reducers if the water pressure is much over 40 psi. Be sure that there are no pipe joints under concrete. Use plastic sleeves around copper pipes where they protrude through a floor slab. Make pipes accessible wherever possible. Always flush pipes before final hookup to fixtures. Pressure-test pipe runs that will be hidden, prior to covering them up.

Vents passing through concrete should be iron. Keep roof penetrations as few in number as possible to reduce the number of possible leak sites.

Heating Systems

In most parts of the country, even the most energy-efficient solar-heated houses have a need for a back-up or auxiliary heating system. Underground houses are no exception to this rule. The principal difference between the heating plant of an underground house and that of a conventional house is size. Since the back-up heating load for an underground house is just a fraction of that of a conventional house with the same heated area, "underground" heating systems are correspondingly smaller. What size should the system be? That, of course, varies with the climate and with how well the house has been designed and built for energy efficiency. But an efficient underground solar house in middle America will require only about 2 Btu per square foot of floor area per heating degree-day ($Btu/ft^2/HDD$), which compares with conventional houses that need anywhere from 8 to as many as 20 $Btu/ft^2/HDD$. So, if you live in a climate where the design temperature is 0°F, then for that design-day there will be 65 heating degree-days. If you have a 1,500-square-foot home, then $2 \times 1,500 \times 65 = 195,000$ Btu per day is the total heating load. The hourly load is $195,000 \div 24 = 8,125$ Btu, which means that a furnace rating of 8,125 Btu per hour is needed. Actually, you probably won't even need that much unless the average outside (ambient) temperature stays at 0°F for several sunless days. This is because the immense mass of the house will take many days to cool off significantly. But, let's assume that the 8,125 Btu per hour is right. Typical residential gas furnaces have outputs in the range of 90,000 to 135,000 Btu per hour, which is clearly much too big for the needs of an underground house. If you were to install such a unit, the system would be very inefficient. But where do you find a whole-house furnace with such a small 10,000-Btu-per-hour rating? As far as I can tell, you can't. Some small mobile-home furnaces are close, but really are not designed for the relatively long duct runs of a full-size house. I recommend not using a furnace at all. With such a low heating load, individual electric room heaters become not only practical but more effective than a central heating plant. If you are going to install a ducting system for central air conditioning, then consider using a conventional gas or electric water heater connected to a fan-coil unit in the ducting system. A 50-gallon water heater will easily provide 10,000 Btu per hour with considerably more cost-effectiveness than a 100,000-Btu-per-hour boiler or furnace.

Because of the high mass usually available in underground houses, their temperatures change slowly. For that reason, the water heater sometimes can be used to heat both domestic water and the living space. When used for domestic water heating, the unit will not be available to heat the house, but it can recover fast enough to be back in the running before the house can cool much below the thermostat set-point. Also, the maximum usage for heating is late at night when little domestic hot water is needed. A small heat pump can be an effective heating and cooling system if ducting is installed. Units in the 1- to 2-ton size range should fit the 10,000-Btu-per-hour requirement quite handily.

For many underground houses, a few well-placed electric radiant or baseboard connective heaters will do the trick with a very small capital investment, and because they are used very little, their inefficiency as compared to other heating systems is not objectionable, nor is the relatively high cost of electric power. Keep in mind that this design size is based on a 0°F *average*

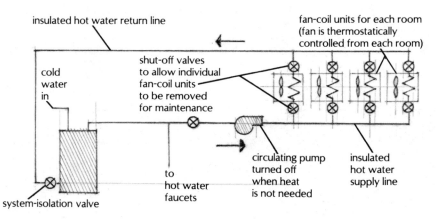

Figure 7-7: For many energy-efficient underground houses, an oversized water heater can provide enough heat for the house and domestic hot water. This illustration shows such a system.

temperature. In most parts of the country, such a day is rare. For 10 to 11 months of the year, the chances are that you will have to dump excess heat from the house — heat that has been generated internally by cooking, water heating and a wild party or two. These heat inputs are called *internal gains*. In an underground house, a housewarming really warms the house, since each guest is contributing about 400 Btu per hour toward heating your house — not an insignificant amount!

Of course, not every underground house is going to be efficient enough to need only 2 Btu per square foot per degree-day rating. But even with twice the load, 4 Btu/ft^2/HDD, the design load for the 1,500-square-foot underground house is still only 16,250 Btu per hour and still not more than the output of a standard home water heater. If the house is very large (underground houses of 6,000 square feet and more have been built), a standard-size furnace can be used, but quite frankly, the idea of intruding on the blessed quiet of an underground home with a furnace fan doesn't seem like a good idea to me. Instead, use individual room heaters for better control and cost-effectiveness.

In any case, be sure to figure the heat loading for your house so that any heating units you purchase are properly sized. The range of heating loads for underground houses seems to be from a micro-load of around 0.5 Btu/ft^2/HDD ("Put out the dog, Ma. It's hot in here!") to as much as 10 Btu/ft^2/HDD which is much like conventional aboveground housing. With a possible range like that, some calculation seems appropriate (see Appendix 1).

Wood Heat

If it's available, a woodstove is my first choice for a heating system. After all, for at least nine months out of the year in most states, a well-designed and well-built underground house is heated by the primary heating system: the sun, the earth and the heat that comes from internal gains. For those three months when just a little help is needed, a wood heater does fine. If your past experiences with wood heat relate to some drafty farmhouse and the gaping maw of a woodstove gulping tons of wood, let me reassure you. Most of the underground houses in my region, the Midwest, use wood for back-up heating. Most that I know of use no more than one cord per heating season, and many of them use a half cord or less. Since a half cord is about the amount that falls out of the trees in a small woodlot, you may never even have to cut a tree!

In order to keep your wood consumption low, you must do one thing that most homeowners ignore: Get a small stove. Until you have heated with wood in an underground home, you probably will not believe me when I say that there are very few woodstoves on the market today that are small enough. Remember, you are probably looking for an output of only 10,000 to 20,000 Btu per hour at the maximum. If you choose too large a stove, you will be plagued with overheating problems, and if the unit can be made properly airtight, you will suffer from excessive smoke and rapid creosote buildup. A properly sized stove will allow you to build a small but adequate fire that burns steadily but not fiercely and does not have to be throttled down to the point where it is literally gasping

Figure 7–8: An outside combustion air inlet represents a potential source of heat loss to the outside if improperly placed. Where possible, the inlet level should be lower than the outlet to the stove. This discourages cold air from convecting into the house when the fire is not burning. Also, if the inlet slopes away from the house, any moisture that collects will drain away from the house.

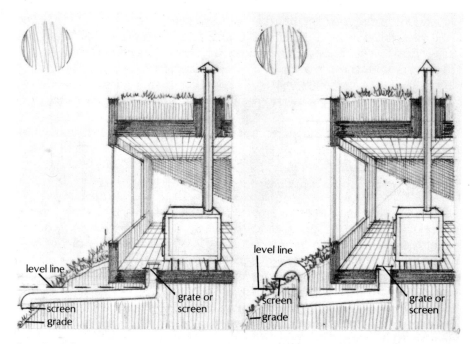

for air. Of course, an important side effect is that the combustion efficiency is much higher, and you actually use less wood for the equivalent heat generated. For most houses, you can ignore wood furnaces; they are simply too big. The small European airtight stoves built for smaller energy-efficient houses are a good choice. Depending on the results of the heating-load calculation, you'll probably find that it's safe to buy the smallest one available.

Notice that I have not even mentioned fireplaces. Nor will I, at least in a favorable sense. As pretty as fireplaces can be, they are rarely useful as heaters. They're hard to control, and you must always build an excessively large fire, since the open hearth needs more burning surface than an enclosed firebox fire. Above all, they are terribly inefficient when compared with a good airtight stove. If you must look at the flames, get a stove with glass doors, not a fireplace.

While most stove manufacturers do not state that outside combustion air is needed for small airtight stoves, I still recommend installing a pipe to the outside for combustion air. Usually the underground house is so tight that it's really poor practice to burn anything (gas, oil or wood) using interior air. Of course, if you have a forced-ventilation heat-exchanger system, then you are moving plenty of air in and out of the house, and combustion air is hardly a problem. If the house is tight, however, there can be problems. The example of one super-tight house comes to mind. When the stove was burning merrily, air would be sucked down through the water-heater vent and would blow out the pilot light. Likewise, if the water heater happened to be burning when the fire was being lighted, the draft would be so poor in the stove as to make it almost impossible to fire up until the

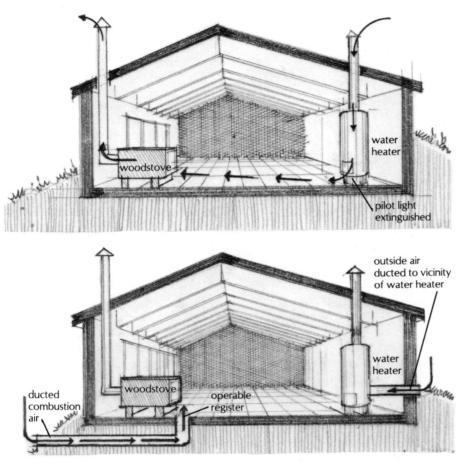

Figure 7-9: In a very tight house, not providing outside combustion air for a woodstove may produce some unexpected results — such as extinguished pilot lights on furnaces or water heaters. The solution is to duct outside air to the vicinity of the woodstove and devices with pilot lights so that there are separate sources of combustion air.

water-heater flame shut down. A source of outside combustion air for both the water heater and the stove was clearly indicated.

Heat Distribution

Because of the low losses from underground houses, heat distribution is not as much of a problem as in less efficient houses. In the case of a wood stove, if room doors are left open, usually enough heat will move through the house to maintain comfort levels. If the house is extraordinarily long and the stove is more than about 30 feet from any room, a small (100 to 200 cfm) fan can be used to pull warm air down the hall through a duct in a furred-down ceiling to wall vents in each room. The end of the furred-down area is at the room with the stove, and exits are placed in each room off the hall. Doors should be left open, since a return air path is necessary. Another way to ensure return air flow is to place a

Figure 7-10: A hallway ceiling can be furred down to make a main air distribution duct for ventilation, heating or air conditioning. Return air registers from each room must be provided near the floor level.

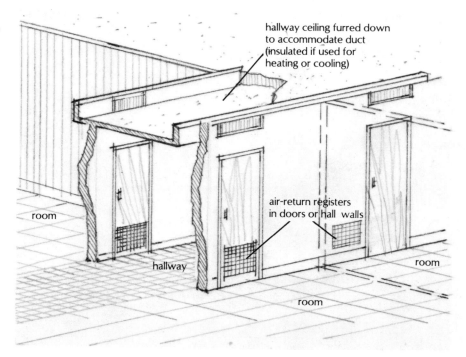

hallway ceiling furred down to accommodate duct (insulated if used for heating or cooling)

air-return registers in doors or hall walls

room

hallway

room

room

register in the bottom part of an interior door or in the lower part of a wall. The fan can be placed anywhere in the duct preceding the first exit register (see figure 7-10). Usually, the farther the fan is from the inlet register, the quieter the system becomes. If the interior of the duct is sound-deadened with lining, then the fan may operate in almost total silence. A manually operated fan is the easiest to install. Simply turn it on when the fire is burning and turn it off when the fire is out. It's OK to leave it on even when there is no fire. The fan can also be thermostatically controlled so that it turns on automatically whenever it senses air above a certain temperature, such as 80°F. You can buy even more complex controllers, which sense not only the temperature of the entering air but the temperature of the rooms at the duct exits as well. I suppose the ultimate is to have a microprocessor-controlled fan with motorized dampers in each room. I have yet to see such a design, but no doubt someone will build one soon. Again, about 90 percent of the time, no fan is needed at all, so I suggest a simple method of control engineering, i.e., an on/off switch.

One area of concern in a woodstove installation is the stack, since it usually must penetrate the roof. Because most waterproofing materials cannot be placed around hot surfaces and be expected to last, a well-insulated, high-quality pass-through is needed. What is probably best is to install a concrete pipe extending above the roof's earth cover. The woodstove stack is then run through that. With poured roofs, the concrete pipe can be set in place on the roof form before the pour. When the concrete hardens, the pipe can be easily waterproofed, and it will not get hot enough to cause failure of the waterproofing. You can place concrete around the metal to cap off the gap between the stack and the concrete sleeve, or a metalsmith can easily provide a water-diverter cap, which can be sealed to the stack and made to fit over the concrete sleeve. Also, don't forget that for a good draft, tall stacks are better than short ones and that the top of the stack should be at least 3 feet above the level of the highest point within 10 feet of the stack.

A word of caution: If your woodstove stack comes through a sod area, don't forget to keep the grass mowed, particularly in the fall, so that there is little likelihood of your roof grass catching fire from sparks. It's a good idea to install spark-arresting screens in any case.

Summary of Heating Systems

Use any of the many available energy-audit systems (see Appendix 1) to calculate the expected heating needs of your home for a winter design-day. If wood in very modest quantities (one cord a year in all but the
(continued on next page)

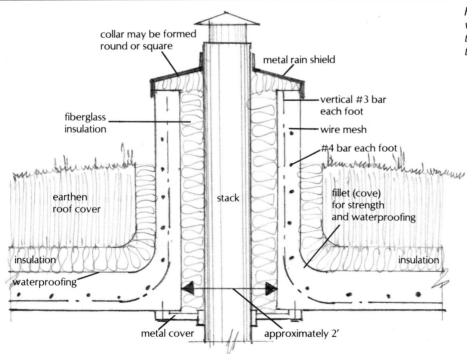

Figure 7-11: Roof penetrations for woodstove stacks are best made through concete collars that extend through the earth cover.

collar may be formed round or square

metal rain shield

fiberglass insulation

vertical #3 bar each foot

wire mesh

#4 bar each foot

earthen roof cover

stack

fillet (cove) for strength and waterproofing

insulation

insulation

waterproofing

metal cover

approximately 2'

Summary of Heating Systems—*Continued*

coldest sites) is available, consider a small, airtight woodstove for a back-up heating system. Always use an external combustion-air source. Use a concrete feed-through for the stack if the roof is concrete.

If a conventional heating system is desired, consider using heaters in each room, sized according to the energy-audit calculations. For a central heating system, a mobile-home-sized unit is about the only type that will be small enough. Be sure to provide both heat-duct entry into each room and a cold-air return to the furnace from each room.

An alternative to conventional furnaces is the use of a small water heater connected as a boiler with one or more fan-coil units used for heat distribution. In mild climates, a heat pump can be a reasonable choice for heating if central air conditioning is to be installed.

Summertime Comfort Control

For most underground homes, humidity control has the most effect on summer comfort. In most of the United States and Canada, it is usually not difficult to maintain comfortable temperatures through night ventilation plus cooling from the earth via the floor and the lower walls. But even at 75°F, humidity hovering near 90 percent is quite uncomfortable. Fans become ineffective because perspiration evaporates slowly at high humidity levels and little cooling is accomplished. The first step in keeping humidity levels low is to reduce or eliminate as many sources of water vapor as you can. For those that cannot be eliminated, try to limit their use to times when outside air can comfortably be circulated through the house to sweep the water vapor out with it. Examples of humidity generators are cooking, particularly boiling and steaming, cleaning household surfaces, doing laundry, bathing and washing dishes. Of course, people emit water vapor, too. In a new underground house (and unfortunately in a few older ones that are poorly damp-proofed), significant humidity is generated by water vapor released from the concrete. Concrete may take many months, some say as long as two years in some climates, to reach minimum moisture levels. Most of the moisture does leave within six months of occupancy, but many people mention a continuing reduction in summer humidity levels for several summers. Part of that, I am sure, is better control over humidity by the occupants, but certainly it is well known that concrete takes a long time to cure thoroughly. Don't be discouraged if the humidity levels of your new concrete underground home are higher than desirable during the first summer. They will be better the next year, and probably better still the year after that. Of course, when you are uncomfortable, you want to do something right away, not just think about how nice it will be in the future.

Other than controlling the entry of water vapor into house air, there are only three ways to reduce humidity levels: by passing the house air over a cold

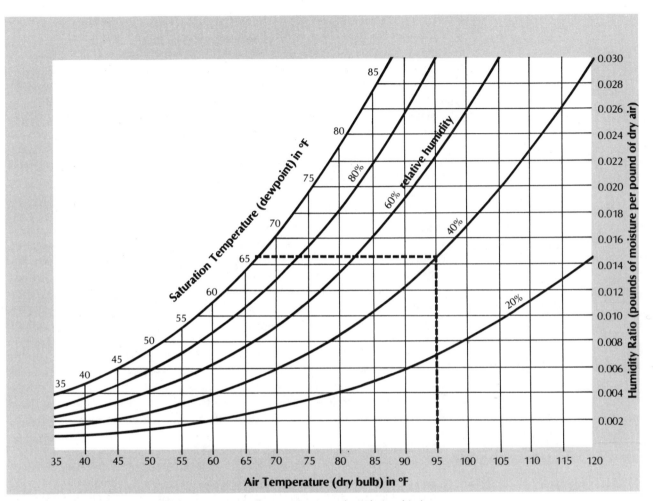

Figure 7–12: This simplified psychrometric chart allows you to see the relationship between saturation temperature (dew point), relative humiditiy, humidity ratio (absolute humidity) and air temperature. The example shows how to determine the saturation temperature (the temperature at which condensation occurs) for 95°F air at 40 percent relative humidity. Find the 95°F air temperature on the horizontal axis, and follow that line vertically until the curved 40 percent relative humidity line is intercepted. Then move horizontally to the left along a constant humidity ratio line until the saturation temperature line is reached. Then just read the saturation temperature which for this example is about 66°F. When temperatures in an underground house fall below the saturation temperature, condensation will occur.

SOURCE: Redrawn and abridged from American Society of Heating, Refrigerating and Air-Conditioning Engineers, *ASHRAE Handbook of Fundamentals,* copyright © 1981, with the permission of ASHRAE, Atlanta, Ga.

surface and condensing some of the water out of the air; by passing drier outside air through the house and sweeping moist air out; and by absorbing the excess moisture with some kind of desiccant. Although experimental desiccant systems are in operation, they are not yet available commercially.

Air conditioners and dehumidifiers operate on the principle of passing room air over a cold surface and condensing out excess water. The only difference between an air conditioner and a dehumidifier is that the air conditioner passes the cooled air on to the room, while the dehumidifier cools the air, condenses out the water, then heats it back up to room temperature (actually a little above). This allows the dehumidifier to operate a little more efficiently, since the waste heat from the dehumidifying process is used to reheat the air. Dehumidifiers cost less than air conditioners, and they're also cheaper to run. For most underground houses, the dehumidification in the daytime with ventilation at night is an adequate strategy for handling comfort problems in the summer.

The second method, ventilation, may be all that is required in some climates. Coastal climates, in particular, allow comfort control through ventilation day and night. Many climates allow ventilation to be the principal comfort control at night and during the day in early fall and late spring. The last thing you want, however, is to have 100°F outside air blowing through your underground home, adding heat to all that concrete mass and guaranteeing discomfort at bedtime. When the outside temperature is above 80°F, don't ventilate. When the outside temperature is below about 75°F, move outside air through as fast as you can to cool down the concrete mass. In many summer climates, keeping moisture from being added to the air during the day and whole-house ventilation at night will do the job. It should rarely, if ever, be necessary to use an air conditioner, except in southern latitudes. Of course, this depends on the degree of comfort you and your family desire.

If you feel that the regimen described above is not adequate for your comfort, first try dehumidification. If you can rent or borrow a unit for a week or so, give it a try before installing an air conditioner. Usually, simply lowering the house humidity will do the job, even at the expense of 1 or 2°F of increased room temperature. A dehumidifier is usually quite a bit cheaper both to install and to run than an air conditioner.

If you really do need an air conditioner, consider using it only for those spaces normally occupied during the day and using night air ventilation for the bedrooms. If night air is not cool enough in your climate, whole-house air conditioning may be the only answer. Anyway, a *much* smaller unit is needed for your underground house than for an aboveground house of similar size. A 1-ton unit will often take care of a three-bedroom underground house. One ton translates to 12,000 Btu per hour of cooling (the amount of heat absorbed in one day by a ton of ice melting). If you choose a whole-house air conditioner, then ducts will have to be installed, although they will not have to be as large or as elaborate as those found in most conventional homes. Because many underground houses are linear in design — that is, the rooms are arranged in a line — a furred-down hallway ceiling may provide enough space for all the ducting.

One option that is considerably cheaper than a whole-house air-conditioning system is the strategic placement of window air conditioners. The average power usage can be lower, and the cooling can easily be directed to just where the occupants are. The disadvantages are noise, appearance in some cases and no controlled air circulation throughout other parts of the house.

For those climatic regions with hot, humid summers and cold but not frigid winters, a small heat pump may be the best year-round system. Houston, for example, has such a climate, and heat pumps have performed well in underground houses there. For climates where temperatures are consistently in the teens and below, heat pumps are not very efficient for heating (many models switch to resistance heating when the temperature falls much below freezing). In those climates, summer cooling and dehumidification needs are usually less, too, which tends to obviate the need for a heat pump.

Cool Tubes and Solar Chimneys

What could be more natural to cool an underground house than air passing through pipes buried in the cool earth? By drawing on our knowledge of the temperatures of soil at various depths, it is easy to see one of the problems: In southern sites, which most need cooling, soil temperatures near the surface are not much below that of room temperatures. By late summer, when cooling loads are quite high, the temperature at 3 or 4 feet has risen to levels that will provide little cooling effect via cool tubes. If the pipes are buried at 8 to 10 feet (or to the constant-temperature depth), soil temperatures usually will allow cooling throughout the warm months. Of course, burying the tubes that deep is pretty expensive unless they can be installed at the same time some other excavation project is underway.

The physical dimensions of the cool-tube system depend on the soil type and moisture content. The better the heat conductivity of the soil, the better the tubes will work. This implies that dense, moist soil is to be preferred. Loose, sandy, dry soil will be poorer. Other factors are the surface area of the tubing that is in contact with the soil and the conductivity of both the tubing and the soil. At this time, not enough data has been gathered from actual installations to allow for generalization on design or performance. The best I can tell, from systems I have seen and from published reports, is that you'll need at least 100 feet of 4- to 8-inch-diameter tubing and no more than 600 or 700 feet. The cooler the earth, the shorter the pipes have to be. So excavation costs are a trade-off between length of run and depth of trench. Of course, it usually is of no particular value to go deeper than the constant-temperature depth unless you happen to penetrate to a soil layer with high thermal conductivity.

General comments about cool-tube installations would not be complete without consideration of drainage. Slopes and drain points have to be established

Table 7-2 _____

Matching Cooling Systems to Climates

SYSTEM	IDEAL CLIMATE
Air conditioning	Climates with temperature regularly exceeding 85°F with humidities above 70%; climates with summer daytime temperatures generally above 85°F and with nighttime temperature drops of less than 20°F
Conduction cooling by contact with earth through walls and floors	Earth temperature of less than 70°F and either dehumidification or climate humidities of less than 30% when temperatures are above 90°F outside
Cool tubes	Climates with summer soil temperatures at cool-tube depth below 60°F; humidities below 50% for outside temperatures between 80 and 90°F and less than 35% for higher temperatures
Daytime ventilation with fans or solar chimneys	Daytime temperatures 85°F or less
Evaporative cooling	Climates with humidities less than 30% and temperatures from 80 to 95°F; above 95°F, humidities need to be less than 20% for effective cooling
Night ventilation with thermal storage	Night temperatures consistently below 70°F for at least 4 hours

NOTE: Using combinations of cooling systems is possible when the summer weather is similar to that of the "ideal climate." But, the use of ventilating-type systems (such as cool tubes and solar chimneys) is unsuitable if dehumidifcation is needed, since ventilation continually brings in moisture from the outside, and heavy loads will fall on dehumidifiers or air conditioners used at the same time as ventilation systems. Conductive cooling or thermal storage with night ventilation can be effective in combination with dehumidification or air conditioning, however.

to ensure that the tubes do not fill with water, which could drain into the house from the surrounding soil. Certainly the best way is to have the pipes level or down-sloped and to have them exit through the side of an embankment. If that design is not possible, the pipes can slope away from the house for some distance before rising again to the surface. At the lowest point, the pipes can drain into a gravel-filled pit, called a dry well, which must always be below the level of the pipe where it enters the house and above the highest groundwater level. Another approach is to use pipe with no drains, installed so that no surface or groundwater can enter the system. Condensation can close off a pipe at an abrupt bend if air humidity is high and the earth is much cooler than surface temperatures (see figure 7-13).

Trial and error is the best way to determine rates of air flow through the tubes unless other systems have been installed and operated locally under similar

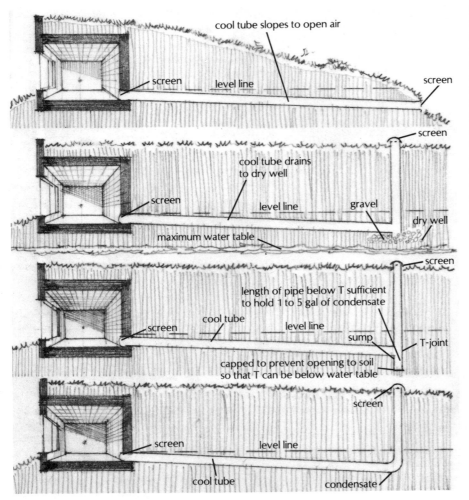

cool tube slopes to open air

screen

level line

screen

screen

cool tube drains
to dry well

screen

level line

gravel

dry well

maximum water table

screen

length of pipe below T sufficient
to hold 1 to 5 gal of condensate

screen   cool tube   level line

sump

T-joint

capped to prevent opening to soil
so that T can be below water table

screen

screen   level line

cool tube   condensate

Figure 7–13: In a cool-tube installation, provisions for drainage condensation away from a house include the following: sloping the cool tube away from the house to open air; allowing condensate to collect in a dry well if the water table never rises above the dry well level (which is common in desert areas); collecting condensate in a sump made from cool-tube pipe and a T-joint; allowing condensate to collect in the cool tube if the tube is 6 inches in diameter or larger and if the air is very dry so little condensation occurs.

soil and temperature conditions. To determine flow rates, use a variable-speed fan. Start by running the fan on its slowest useful speed, and measure the outlet temperature on a hot day. Increase the speed each hour until the outlet temperature starts to rise. Leave the fan at that speed setting for a day or more. If the temperature of the outlet air rises more than a few more degrees, slow down the fan. Keep raising and lowering the speed in increments over a number of days until a stable outlet temperature below 70°F is reached. That speed will be about optimum. You will find that the optimum speed may vary over the summer if the pipes are above the constant-temperature depth, since the earth temperature will be rising over the summer months. That means your cooling capacity (and the optimum fan speed) will be falling as the summer progresses.

Here are a few precautions to note when installing a cool-tube system. It's a good idea to place wire screen at both ends of the pipe. Critters and insects like cool, dark places, too. Also, choose the entrance site carefully, preferably in a shady spot where inlet air is cooler. You will not appreciate drawing air from the barnyard or from your sewage lagoon into the house, cool or not.

A Solar Chimney

The solar chimney is a device that uses wind and solar energy to create an updraft in a chimney. The usual method is to build a wooden chimney structure with the south, and sometimes the east and west, surfaces glazed. Inside that part of the chimney there is usually a black metal surface to absorb the sunlight and create a hot surface. The heated air rises and draws the warmest air out of the house. The presence of wind can augment the solar-induced draft. In climates where the summer daytime air temperatures are not far above the comfort range, this method can provide all the ventilation a house needs. Even at night, if the inside of the house is warmer than the outside air, a significant amount of ventilation is created by a solar chimney. On the other hand, in most summer climates, the maximum rate of ventilation occurs when the outside air temperatures are near their maximum levels — usually far above a comfortable temperature — and drawing in outside air is not a good idea. Forced-air ventilation or wind-powered ventilation that can operate at night when it is cooler makes more

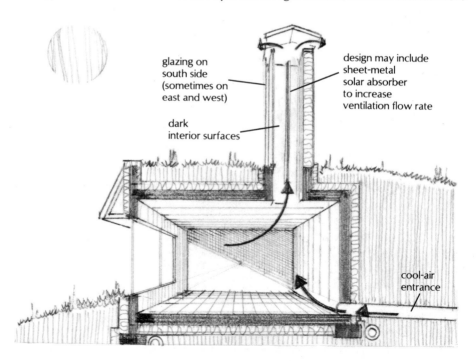

glazing on
south side
(sometimes on
east and west)

design may include
sheet-metal
solar absorber
to increase
ventilation flow rate

dark
interior surfaces

cool-air
entrance

Figure 7-14: In some climates, solar chimneys, which draw heated air up and out of a house, provide good ventilation.

sense. To be honest, most ventilating chimneys billed as solar chimneys have such a small relative area of solar collection surface that most of the ventilation added by the presence of the chimney occurs because of the wind and the rising of warm air from inside the house. I have not been able to get very excited about ventilating chimneys, except for a few sites. Those sites either have few days when it is too hot to ventilate with outside air or the site has a source of cool air that remains significantly below the outside ambient — such as that provided by cool tubes. In Missouri, which is often called the cave state due to the numerous limestone caverns there, several houses draw air for ventilation from caves. The solar chimney can do a good job in these cases and can provide truly natural air conditioning at modest costs.

Ventilation

While solar chimneys and cool tubes provide ventilation as a by-product, cooling is their primary purpose. Ventilation does indeed help with cooling, though its primary purpose in underground houses is to exhaust stale house air and introduce fresh outside air. However, in terms of energy conservation, it is not always to the homeowner's advantage to exchange inside air with hot or cold outside air, no matter how fresh it is. In most houses, there is no choice, because the house is not tight enough for interior air pollution to be a problem. Outside air enters continuously (infiltration) and without owner control at rates that are usually higher than those necessary to maintain fresh air in the home. In the underground house, infiltration is naturally going to be much lower than in just about any other type of house. A conscious effort must be made to maintain a fresh, pleasant, unpolluted interior environment by having a formal exhaust-intake ventilation system that can do more than just exchange indoor air with outdoor air.

First of all, what pollutants are involved? There are three distinct classifications: chemical pollutants, particle pollutants and nuclear pollutants. Chemical pollutants cause odors or are harmful to the body's metabolic system. They range from the benign but unpleasant odor of garbage to the not-so-benign but nonodoriferous carbon monoxide, which may come from a gas heater, cook stove or water heater. Smoke, dust, bacteria and pollen are particle pollutants. Nuclear pollutants, most notably radon, are naturally formed from the minute amounts of radioactive materials found in the earth. I hesitate to make nuclear pollutants a separate class, but since there has been a lot of discussion recently about radon and its probable presence in super-tight houses, I have included it. As far as I can tell, there is as yet no reason to believe that health problems will be caused by the amount of radon that can be found in a house, but in response to the concern raised by others, it should be considered.

The goal of ventilation is to maintain indoor air that is essentially free of pollutants, and there are two basic strategies for achieving that end. The best (and perhaps most difficult) is to arrange a house and life-style to minimize the creation of pollutants. When they are created, the second approach is to use ventilation to remove from the air the pollutants that are already in the house.

Sources of Indoor Pollutants

Just about every room in a house harbors a pollutant generator of some sort. The kitchen, however, is often the worst offender. Cooking odors, smoke, combustion residues from gas flames, and odors arising from the storage of foods and food products all can be problems. Cleaning chemicals can be powerful producers of chemical pollution. In a few cases, they can even produce very toxic gases. For example, ammonia and chlorine bleach are bad enough separately, but when mixed, the result is really unpleasant. Bathrooms are not particularly bad polluters, although their distinctive odors do add to the overall problem. Other rooms of the house can add to indoor air pollution. Vinyl, carpeting, draperies, furniture made largely of particle board (which is just about all of today's products) and even your friendly woodstove may add compounds to the air, such as formaldehyde or carbon monoxide. And, of course, the ultimate indoor polluter is a smoker.

If it all sounds like a big problem, you should know that even in the underground house, there are few incidents of *known* health problems resulting from indoor pollution; the important exception is carbon monoxide. What is not known are the long-term effects and the psychological effects of living in a tainted atmosphere. I don't know, and I don't want to be a test case, so my recommendation is to maintain a fresh-air indoor environment.

Artificial Respiration

Back in the dim past when I was a tenderfoot in the Boy Scouts, I learned that to time the cycle of artificial respiration I had to repeat continually, "Out with the bad air, in with the good." That is what ventilation is all about. The idea is simple, but as usual, doing the job correctly requires careful thought and design. First of all, how much ventilation is needed? It is generally accepted that the total volume of air in a typical house needs about one exchange with outside air every one to two hours. One exchange every two hours equals 0.5 *air changes per hour* (ac/hr). With systems in the house to clean and purify the air (like lots of plants), fewer air changes per hour will not be detrimental to comfort or health. On the other hand, in a houseful of cabbage-eating smokers who suffer from flatulence, more air changes may be wise!

In most houses, the air-exchange rate is more than adequate due to uncontrolled infiltration through cracks around windows and doors, through gaps between the floor and walls and through other sites of loose construction. In an underground house, air paths between inside and outside are much fewer, and with energy-conscious occupants, few cracks will be left uncaulked and few doors un-weather-stripped. In a tight house, the naturally occurring exchange rate of 0.25 air changes per hour or less is probably not enough for maintaining good air quality and comfort. Comfort suffers because such low air-change rates mean that humidity can be significantly increased along with odors. If there are gas-fired or wood-burning appliances within the living space, air quality can deteriorate to the point of being a health hazard. But having a super-tight house is not a disadvantage; quite the contrary, it's a tremendous advantage when it comes to ventilation.

While not-so-tight houses may not need a specific ventilation subsystem, they allow little control over ventilation rates or incoming air temperatures. However, you can have both through the use of an air-to-air *heat exchanger,* also called an *economizer.* In a very tight house, like a well-built underground house, ventilating air is drawn in from the outside with a small fan at a rate consistent with meeting the goal of one air change at least every two hours (0.5 ac/hr). While the air is coming in, air must be leaving, too, or the house would swell up and burst (well, not really). In the economizer system, the air leaving passes through a heat exchanger and passes most of its heat to the incoming air. On a cold winter's day, the warm air leaving the house heats the cold air entering. On a hot summer's day, the cool air leaving cools the incoming hot air. The end result is a system that allows as much ventilation as we need or want with little energy penalty in space heating or cooling.

Economizers have been around for a long time, but unless they are connected to a very tight house they are useless. In a conventional house, if air is drawn from the outside through the unit, warm air exits all over the leaky surface of the house and little goes out through the economizer. In the tight house, of course, most will leave through the economizer. In fact, one way to see how well the weather stripping and caulking are holding up on an economizer-equipped underground house is to compare the temperature of the air coming into the house with the temperature of the air already in the house on days with comparable outside temperatures, but months or years apart. Let's consider a day when the outside temperature was 40°F and the inside temperature was 70°F. If the ventilating air temperature from the economizer was 65°F when the house was new but 50°F after two years, you may be in for a new round of caulking and weather stripping. Of course, you might be having filter or economizer problems, too.

Commercial economizer units are generally sized according to the volume of the space to be ventilated. Home-built units should be also. In most houses, all you have to do to calculate the volume is to multiply the total floor area by the ceiling height. If you have a dome or sloped ceilings, it is a little more complicated, but the same principle applies. For example, if your 1,000-square-foot house has an 8-foot ceiling, its volume if 8,000 cubic feet. Changing that volume of air each hour requires an economizer fan that will move 8,000 cubic feet per hour or, as fans are usually rated, about 133 cubic feet per minute (cfm). This is small as fans go, so the power used is minimal, particularly in relation to the potential

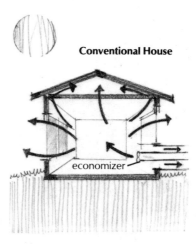

Conventional House

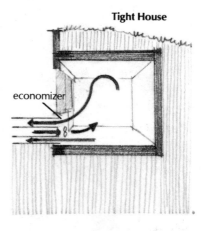

Tight House

Figure 7–15: An air-to-air heat exchanger, or economizer, works well in a tight house, since the primary exit for air is through this unit. A conventional house loses air at many sites, which means that there would be too little outgoing air through an economizer to preheat incoming air.

improvement in comfort and the energy savings in preheating and precooling intake air. Typically, these units cost less than a room-size dehumidifier, yet over a summer they may remove more excess water vapor than the dehumidifier could, and at less cost for electric power.

Note, however, that economizers neither dehumidify in the summer nor humidify in the winter, although they do reduce the amounts of water vapor added by household activities. If the air outside is hot and humid, the air coming in via the economizer will be cooler, but will contain just as much water vapor. This may mean that you may wish to reduce the ventilation rate in the summer if you are air-conditioning or dehumidifying house air. If you are not, you may want to consider actually increasing the ventilation rate, particularly when you are doing lots of laundry, scrubbing floors, entertaining a troop of active Boy Scouts, or participating in any other activity that dumps additional moisture into the air.

Distribution of Ventilating Air

According to the complexity of the floor plan, needs will range from very simple systems to fairly complicated approaches. One approach is a design that delivers the ventilating air to the main living space and picks up the outgoing air from the kitchen. Then, to ensure that the rest of the house gets its share, a circulator system mounted on the hall ceiling can pick up the air and pass it through the other rooms. Frankly, such complexity is rarely needed, since the main source of interior pollutants is the kitchen, with the living room close behind. The same low-speed circulating system that keeps heat (or coolth) distributed evenly through the house does nicely for keeping fresh air evenly distributed, although if it is being used only for fresh-air distribution, the flow rates can be even lower than those used for heat distribution.

Air Cleaners

Airborne dust can be filtered out of room air fairly easily. Most interior circulating sytems associated with air conditioning and forced-air heating include filtration. If these filters are kept clean, by replacing throwaway units or washing and properly treating permanent ones, removal of dust is largely taken care of. Filtering out smoke, pollen and bacteria, however, is more difficult. Filters that do a good job of straining out these very fine particles usually require fans that are considerably more powerful than those found in heating or air-conditioning systems. For home use, *electrostatic precipitators* do a good job. These electronic air cleaners place an electrical charge on the tiny particles, and before they can get away, an opposite charge attracts them into a trap, allowing only clean air to leave. Versions that mount in a furnace or air-conditioning duct are available, as are free-standing units designed for a single room. Their power consumption is not large (although the units themselves are relatively expensive), and they can dramatically reduce the amount of wall cleaning you have to do, even in houses with relatively clean air. Years ago, room air filters that used wet or chemically treated cloth surfaces were used, but their effectiveness was

*Photo 7–3: An air purifier such as this one with a
HEPA (high-efficiency particulate air) filter is effective
for cleaning room air.*

marginal, except for large-particle dust, and the wet units also acted as humidi-
fiers, which are usually not welcome in underground houses.

Another air-cleaning device that functions well is one that contains a HEPA
(high-efficiency particulate air) filter. Such devices, which were used for years in
industrial, medical and pharmaceutical applications, have been available for
residential use only within the last few years. Portable room-size units or
whole-house units that are installed adjacent to heating and cooling systems are
now available. Air Techniques, Inc., of Baltimore, which manufactures both
types, claims that its Cleanaire units have a 99.99 percent filtration efficiency for
particles 0.3 micrometer (which is 1/75,000 of an inch) or larger.

Air Deodorizers

So-called room deodorizers that rely on a chemical to mask odors may
become a part of the air-quality problem rather than a solution. They work in one
of two ways: The scent from the deodorizer overpowers the unpleasant odor to
the point where it is no longer sensed, or the deodorizer chemical desensitizes
your sense of smell so that you simply can't smell the offending odor any longer.
In both cases, the odor is still there, but you just can't smell it. For a guest just
entering a house, a little time is needed for the desensitizing to occur, and the full
force of both the original odor and the masking agent assaults the guest's nose.
What was that about first impressions?

The kind of deodorizer I recommend actually absorbs the odor-producing
chemical and removes it from the air as an electronic precipitator removes dust.
Most of these odor-absorbing units have activated charcoal as the absorber, and
while I have yet to see a room-size or home-size activated-charcoal deodorizing
unit, electronic air-cleaning units often include a charcoal filter for odors. Ductless
kitchen stove hoods have no vent to the outside but rather circulate kitchen air
through a layer of activated charcoal. These hoods are specifically intended to
take care of odors given off from the stove, but anytime they are operating, they
are also deodorizing the air in the vicinity. They have a limited capacity for

absorbing smoke and for condensing and absorbing grease, but they are not as effective as electronic precipitators or vented hoods for that purpose. Of course, the vented units rely on the "Out with bad air, in with the good" approach rather than absorbing the pollutants.

A deodorizer that is common in many homes is the houseplant. Plants breathe through their leaves, and in the process, odors are absorbed. Naturally, the number and size of the leaves on a plant have an effect on its capacity. A greenhouse full of plants will do a very good job of freshening a houseful of air if the air is circulated fairly continuously through the plant area. Of course, some plants may contribute to odor problems. Not everyone believes that a greenhouse full of tomato plants is a pleasure to the nose.

What about Radon?

Both sensational news accounts and staid scientific journals have considered the radioactive gas radon as a potential, if not actual, health hazard in super-tight houses with very low ventilation rates (below 0.5 ac/hr). They refer in particular to basements and underground dwellings. It seems that radon is a product of naturally occurring radioactive minerals commonly found in trace quantities in earth and igneous or metamorphic rock. Houses built of granite or made with concrete that includes nonsedimentary rock have higher concentrations of radon than wood-frame, aboveground houses. Also, basements or below-grade houses will have increased concentrations because radon is heavier than air and thus settles to the lowest parts of a structure. It must be realized, though, that the concentrations we are talking about are very slight and occur only if there is virtually no air exchange with the outside. Even in the tightest of houses, there is enough air exchange due to occupant activity to disperse the minute amounts of radon that might accumulate. I cannot get alarmed about these amounts, but the only way to be sure is to provide some controlled air exchange with the outdoors.

Designing Ventilation Systems

The optimum design of a ventilating system for an underground house is dependent on the needs of the owner, the climate and the house design. The owner may require air that is free of pollen because of allergy problems. Outside air may be polluted, and minimal exchange with the outside is desirable. In a climate characterized by long, hot and humid summers, ventilating air is not an advantage, because of its moisture content. Air passing through an economizer will still retain its moisture, which will result in a need for increased air conditioning or dehumidification to lower humidity. Under any of those conditions, air exchange should be minimized in favor of more interior air purification.

On the other hand, in a climate with mild summers and clean ambient air, increased air exchange between the house and the outdoors is by far the most cost-effective approach. If winters include more than 2,000 heating degree-days, an economizer should be considered in the ventilating package.

For the climate that has pleasant nighttime conditions in summer but hot, humid days, a combination of ventilation and purification methods may be best. An unfortunate aspect of this in-between climate, which characterizes vast areas of the Midwest and Northeast, is that buying a lot of equipment for a need that may exist only one or two months of the year is not very cost-effective. A better approach is to modify your living cycles so that the need for air purification occurs when ventilation is practical.

By minimizing polluting activities during the heat of the day when ventilation is not appropriate, air quality in the house can stay high enough that adding air-purification equipment will not be necessary. Woodstoves are not going to be used in summer, and the use of gas ovens and range tops can be avoided so that carbon monoxide and other combustion-product buildup should not be a problem. With the addition of a few strategically placed large plants and the exclusion of cleaners that have solvents, ammonia or other strong-smelling ingredients, indoor air quality can stay just fine. All polluting activity can then take place in the evening, when outside temperatures are cool enough to allow for direct ventilation.

Summary of Ventilation Systems

Three basic approaches are usually combined to maintain air quality inside the very tightly built underground house. The cheapest and most effective is simply to avoid activities that pollute. For those pollutive activities such as cooking, cleaning and entertaining, try to choose a time when outside air can be used to clear out interior pollution.

When that isn't possible, some form of air purifier is needed. If particulate pollution is a problem (usually dust or pollen), then the electronic air cleaner is good. If odors are the main problem, circulating house air through an activated-charcoal filter can do the job. Plants are excellent as air purifiers where pollen is not the problem, and they are generally recommended as first-choice air purifiers.

When air exchange with the outside is the best approach for keeping clean air in the house, the tight construction of the underground house allows controlled air entry through purification devices, filters and heat exchangers so that even in sites with polluted air that is much colder (or hotter) than is comfortable, the air that is introduced is clean and odor-free. Flow rates should be such as to move a minimum of one-half to one air exchange per hour. Higher rates of air exchange are not needed for maintaining fresh air, only for cooling when outside air temperatures are in the comfort range.

Lighting

Comfort is very hard to define, but you certainly know when you don't have it. When I sit in a room, I am immediately aware of discomfort if the temperature or humidity is too high or too low. It is obvious to me if the chair is too hard or too soft. But even if the air is just right and the chair perfect, I still may have a clear but indefinable feeling that all is not well. In short, I am uncomfortable. If the skin reports comfort, the ears and nose are happy, and there is nothing unpleasant to taste, what is left? The eyes. The appearance of our surroundings, the way they are illuminated, can aid or degrade comfort. This is graphically illustrated by the difference between the inside of an expensive restaurant and a fast-food corral. In the former, illumination is subdued, contrasts between light and dark are moderate, and boundaries are ill-defined. Fast-food establishments are brightly, even harshly, lighted. Colors are garish and contrasty. It is an environment *designed* to keep the customer tense, ready for flight and on the move. Volume is their business; dawdling over a Big Mac is unthinkable. In a fine restaurant, you are surrounded by illumination cues to make you comfortable, to relax you or to keep you there for an extra glass of wine or dessert. Believe it or not, I, who have no great record of waiting patiently for anything, have been in restaurants where I was actually annoyed if service was fast. Now that's comfort!

A home can have the same supreme comfort by taking cues from the restaurant example. First on the list is *contrast*. Contrast is a measure of the differences in illumination between the light areas and the dark areas of the room. Our perception of contrast is somewhat increased by an increase in sharpness of boundaries between the bright and the dark areas. For relaxing, contrasts should exist, but objects in the darker area should be clearly visible, and there should not be a sharp boundary between light and dark spaces. An example of excessive contrast might be a south-facing window in a dark-colored room — some people's idea of the perfect underground solar room. The space including the window is so bright that the room is too dark for visual comfort. The eye will adjust to the average of the illumination, and if there is too great a difference, it cannot easily see in either the bright or the dark parts of the space. At the opposite extreme is a room that is very evenly illuminated and all decorated in about the same shade of beige. The eye is hard pressed to determine the scale of the room, and the outlines of the contents are vague and ill defined. While both styles of lighting have their place for special tasks, neither is comfortable.

The Wade Theory of Comfortable Illumination says that the illumination found in a forest on a pleasant summer's day is about perfect. Early man was mostly a forest dweller, and we continue to have a love affair with trees. We feel safe in that forest; we feel comfortable. So if we light our living rooms with a source above the head, provide shadowed areas and light areas of diffuse illumination and maybe throw in a few leaves here and there, *voilà*, comfort.

Photo 7-4: A greenhouse not only provides a source of solar heat but also a view, diffuse illumination and humidification.

A common design error in underground houses is allowing extremes of contrast to occur in illumination. The design error is in placing windows on only one side of a room while having dark carpeting and possibly dark walls as well. Having windows only on one side of a room is not a sin; in fact, that is really hard to avoid in truly underground designs. What *is* a design error is allowing those windows to have much higher levels of illumination than the inside of the house. Light-colored carpets, furniture and walls will spread the light out and reduce contrast. Many office buildings have no windows at all, yet they are not considered cavelike. Indeed, if you take the Mammoth Cave tour and have lunch in the Snowball Room, it is hard to believe you are in a cave. Why? It is well and evenly lighted. If you happen to run across one of those underground houses that feels like a cave, come back at night. I will wager that with the sun down and the lower contrast of artificial light, the cave feeling will be gone.

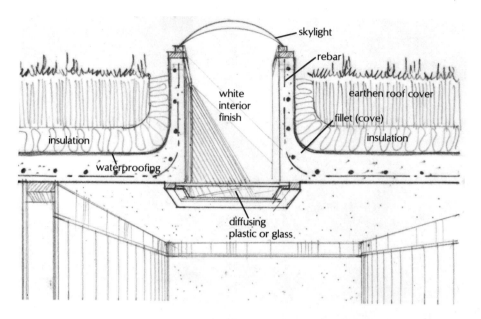

There are several ways to keep the contrast between the windows and the inside at comfortable levels. One is to have a view that is dark in color, or at least not reflective. Looking out on a large white concrete patio is much less comfortable than looking out on a lawn or grove of trees. Interior surfaces should be light in color, particularly in those parts of the room that are not directly illuminated by the sun. A greenhouse or sunroom can be placed between the living room and the outside. The greenhouse diffuses the sunlight and allows more even illumination in the room behind. And, of course, one sure way to keep one-sided illumination from being a problem is to have two-sided illumination with help from a clerestory or monitor-type skylight.

Daylighting

There is considerable art and some science in piping outside light into a house. With windows that aren't used for passive solar space heating, the goal is to have as small an opening to the outside as possible to reduce summer heat gains and winter heat losses and yet still have the light spread over entire rooms. Keep in mind, first of all, that light travels in a straight line. If a skylight is mounted at the top of a light shaft, then the result will be a bright rectangle just below the shaft, with little light diffused throughout the room. At Monticello, Thomas Jefferson, taking a cue from Greek and Roman designs, made his skylights in a cone shape so that all the floor area of the room could "see" the skylight. The result was even, medium-contrast, i.e., comfortable, room illumination. The domed room is ideally suited to illumination from a small skylight near the peak. A *diffusing* skylight is a bright light source, and if the light can reach all parts of the

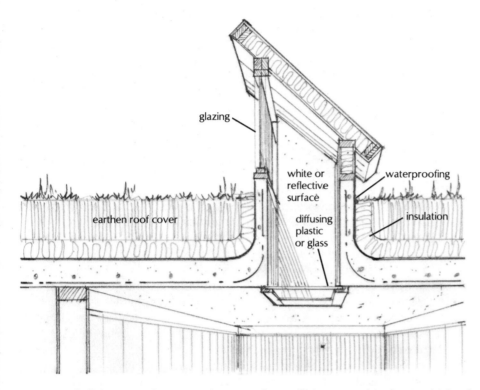

labels: glazing · white or reflective surface · diffusing plastic or glass · earthen roof cover · waterproofing · insulation

Figure 7–17: A monitor or clere-story skylight is the solution to the need for daylighting without in-curring unwanted solar gain in summer.

room, a skylight area of just 1 or 2 square feet will do a surprisingly good job of lighting a room. And such a small skylight is not too damaging from a summer heating or winter cooling point of view. The key word there is diffusing. With a transparent glazing, the skylight will simply pass a sunbeam the size and shape of the skylight into the room. Diffusing glass or plastic glazing will collect light from all over the sky and deliver it all over the room (see figure 7–16).

Even under the best design circumstances, the conventional skylight is a small, high-intensity light source and, at least if one looks up, is a great contrast to the rest of the space. The clerestory or monitor skylight can provide a larger area of light source at a lower contrast. The glazing is vertical, and the effective light source is the sloped or flat surface behind the glass, as shown in figure 7–17. The bright sunlight is allowed to fall on a large white surface, and that surface effectively illuminates the room below by reflection. It becomes sort of a natural indirect-lighting system.

Light shafts can be turned into low-contrast illumination systems by the addition of a diffuser at the ceiling. Placing a small skylight above a wall can turn the wall surface into a light source as in figure 7–18. The principle to be observed is to use a small skylight and cause it to illuminate a much larger surface so that the localized brightness is not too high for comfort.

Atrium areas as sources of illumination can reduce contrast at exterior windows. Some designers have actually painted a north atrium wall white so that the windows opposite that wall would have maximum light. Remember that the eye automatically adjusts to a very wide range of illumination, and in the home, at least, comfort depends not on how bright the light is, but rather on the range of brightness. Having to adjust simultaneously to a very bright and a dark area is not comfortable to the eye. It is much more comfortable to have a view of a low-brightness surface through a window than one designed to reflect lots of light into the room. In fact, if your view is to an atrium, it may be to your advantage to use tinted glass, as long as that glass is not intended for solar gain.

Artificial Lighting

For living spaces, emulate my forest. Have varying illumination across the room, but not abrupt changes between light and dark. As a rule of thumb, when looking at any area of the room, there should be no sense of indistinctness about adjacent areas. There should be no large dark areas (where forest predators can hide . . .) nor very bright areas where you find it hard to see into other parts of the room. Indirect lighting has been used in many underground houses with good effect. Conventional lamps with diffusing shades allow sufficient light for local tasks like reading or sewing, while at the same time spreading light around the rest of the room. The reflector type of lamp is too harsh unless other broader-based room illumination is also provided.

A common design error is placing bright kitchen lighting where it can be seen by occupants in other parts of the living space. The tendency for large open spaces in underground houses is certainly laudable, but it is still good practice to keep bright kitchens lights from being in the direct line of vision of people seated in parts of the house that are intended for relaxation. The same is true for exterior lighting. Lights that are placed for the convenience of arriving guests, for the inconvenience of burglars or simply to show off the house at night are poorly placed if they are directly visible by people inside. Mercury-vapor streetlights are particularly offensive. Sometimes it is simple to design room and window arrangements so that the lights are not a problem, but often the only recourse is to pull the curtains.

As a general design criterion for living spaces, use indirect or low-wattage fixtures for general illumination, with brighter, but still diffuse, illumination for task lighting. Since the brightness of light from a bulb decreases rapidly with distance, task lighting should be close to the task. Although it is impossible to have spotlights some distance away with little loss of brightness, be wary of too great a contrast between the area they light and surrounding space.

For workrooms like kitchens, some designers have tried to use the same system of low general illumination, with task lighting at strategic points. The problem is, every place in a kitchen seems to be a strategic point. One home that comes to mind had all countertops well illuminated with fluorescent strips under

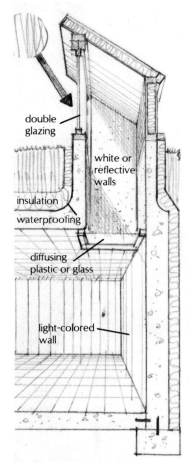

Figure 7-18: A light shaft with a diffuser at ceiling level can illuminate living spaces that are more than 3 feet below grade. A secondary source of illumination, in this case a light-colored wall, provides good light distribution.

double glazing

white or reflective walls

insulation

waterproofing

diffusing plastic or glass

light-colored wall

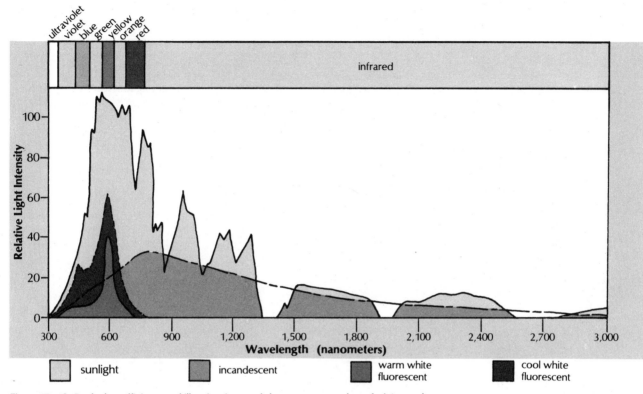

*Figure 7-19: Both the efficiency of illumination and the apparent color of objects change
with the spectral distribution of light from different sources. "Warm" lighting has more
red light, and "cool" has more blue.*

SOURCES:
    1. GTE Sylvania, Inc., *Sylvania Lighting Handbook,* 3d ed. (Danvers, Mass.: GTE Sylvania,
Inc., 1969).
    2. John Kaufman, *IES Lighting Handbook,* 2d ed. (New York: Illuminating Engineering
Society, 1954).
    3. Edward Mazria, *The Passive Solar Energy Book* (Emmaus, Pa: Rodale Press, 1979).

the upper cabinets and indirect lighting at the ceiling. There was plenty of light at
the work surfaces just as the designer planned, but woe to the person who
wanted to find something in the recesses of the cabinets. All in all, it was not very
successful. In kitchens, it is better to have bright but diffuse general lighting. That
is not to say that four 8-foot fluorescents recessed into a ceiling glazed with
diffusing plastic are necessary. That's excessive. Our kitchen is well illuminated by
a single *22½-watt* circular fluorescent bulb located about 6 inches below a white
ceiling! Light colors and white appliances help a lot, but more than 100 watts of
fluorescent illumination should never be needed for the average-size kitchen.

Color should be considered as a part of lighting, too. If comfort is to a great extent psychological in nature, color must be a part of it. Indirect lighting and skylighting, which both depend on reflection from walls or ceilings, will be affected by the color of the light itself (in the case of indirect lighting) and by that of the reflecting surface. If you have a light green ceiling, don't be surprised if your indirect-lighting system gives everyone a sickly look. If your indirect-lighting source is fluorescent, a beige or peach-colored ceiling will keep the room light from having the blue-white character of fluorescent lighting. Tinted or stained glass in skylights can be better than clear or white glass for some areas like entry rooms and hallways.

The color of the light itself varies with the source. Incandescents are generally more mellow than fluorescents, since they include more red light. Some of the new fluorescent fixtures have special bulbs that include an increased percentage of red and yellow phosphors, and they are very similar in appearance to incandescents but much more energy efficient. A lot of people seem to complain about fluorescent illumination being harsh or causing headaches. Some studies seem to indicate that bulbs which have a softer, more incandescentlike coloration are less likely to elicit those complaints. Ten years ago, almost no one would have considered installing a fluorescent fixture in a bedroom. Today, they are not only becoming common, but with circular bulbs, they even look like the familiar incandescent fixture, which was designed to look like the gas light, which was designed to look like the . . .

Summary of Lighting

Whatever the light source, design so that there are no illumination extremes. There should be no difficulty in clearly discerning the details of objects in a room. Any bright light sources should be out of the ordinary range of vision, or they should be compensated for with other bright sources placed around the room.

Skylights should be designed to spread their light over the entire room rather than just under the glazing. Skylight systems using vertical glazing should illuminate ceiling or wall areas several times the area of the sylight to keep contrast at an acceptable level.

Artificial lighting should be equal to the needs of the space but no brighter than necessary. The same rules about contrast between light and dark areas apply to both artificial illumination and daylighting.

Comfort depends more than most people realize on the quality of the illumination in a room. To maintain comfort, the overall level of illumination should be adequate to the task at hand, and most importantly, there should not be too wide a brightness range between light and dark areas in the room.

CHAPTER 8

A Gallery of Underground Houses

Each of the following eight case studies represents different design techniques. Some have worked just as the designer planned; some have turned out a little different from the plan. The important thing to remember is that these houses, which were built in a variety of climates and geographical areas, are *real,* not just lines on a plan. The designs may give you ideas for your house, and you may well benefit from the experiences voiced by owners. A glance at figure 8–1, which shows verified earth structures in North America, indicates that you won't be alone in your desire to live within the shelter of the earth. According to Kathleen Vadnais, editor of the bimonthly *Earth Shelter Living,* "If trends continue, earth sheltering will be included in 50 percent of the new construction in the next 20 years."

Key:
- ● Residential
- ○ Business
- △ School
- □ Agricultural
- * Public
- † Church

Figure 8–1: This map indicates verified earth shelters in North America. Locations are purposely vague to protect the privacy of owners.

SOURCE: Redrawn with permission from *Earth Shelter Living* magazine, copyright © 1982, WEBCO Publishing, Inc., St. Paul, Minn.

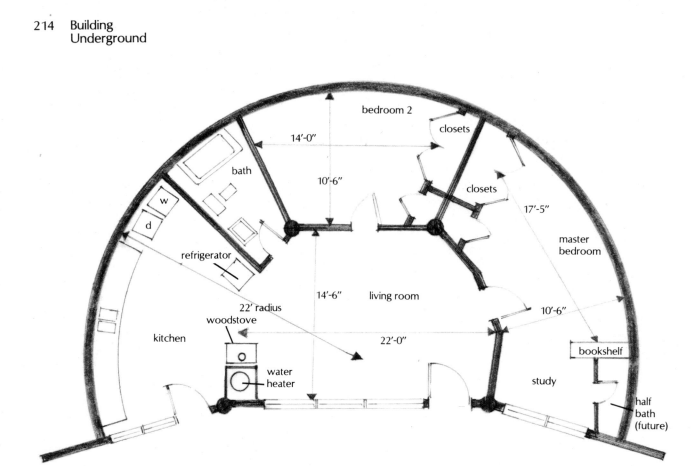

Figure 8–2: The Cooper house floor plan.

The Cooper House

Nestled into a south-facing hill in the eastern United States is a concrete-block and ferrocement house designed by John Cooper and built by Cooper and Richard Potts. The basic house plan is a semicircle with 200 square feet of south-facing glazing, which means that in this 4,800-heating-degree-day climate, the sun provides nearly all of the space heating. With its curved ceilings and free-form surfaces, the house actually seems larger than its actual size of 1,100 square feet. Five feet of the exterior walls are constructed of 8-inch concrete blocks reinforced both horizontally and vertically and grouted full with concrete. Onto these walls, Potts and Cooper constructed a grid of ½-inch steel reinforcing bars covered with metal lath (expanded wire mesh) to form the basis for the 4-inch-thick ferrocement roof section. They decided to build their own spray-concrete application machine, since the cost of commercial equipment was much higher than they could justify for the construction of a single house. Rental equipment was available, but it would have had to travel a great distance to the site, and the cost of building the equipment was a little less than the rental fee.

Since this was their first experience with ferrocement construction, Potts and Cooper built very conservatively. They first troweled on a layer of concrete from the inside and then sprayed on the remaining concrete in layers from the outside. Since it was necessary to keep each of the layers wet between spray applications, the process took ten days to reach the final thickness of 4 inches. While the result was excellent, they believe that the job could have been completed in half the time with no loss in quality. The interior of the walls is finished with a coat of cement plaster, which provides apparently continuous and graceful surfaces. The exterior of the walls is covered with 4½ inches of urethane foam, which is protected from the elements by a fiberglass-reinforced finish coating.

Although the financial institution that granted a mortgage for the house required a conventional heating system, the electric baseboard units that met that requirement have not been used. Instead, a woodstove, designed and built by Potts, provides back-up heat. Of particular interest is the unique air-circulation system in this house. Under the floor there is a network of ducting, which

Figure 8-3: Cutaway view of the Cooper house.

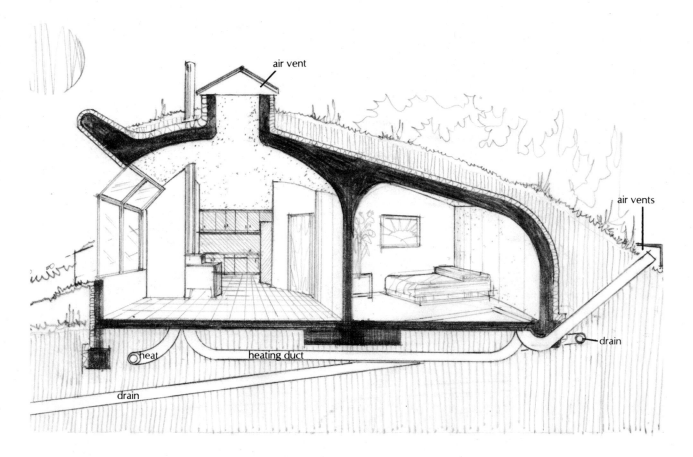

air vent

air vents

drain

heat

heating duct

drain

Photo 8–1: The Cooper house.

terminates in floor registers in each of the rooms and under the woodstove. The rising heated air from the stove draws cool air through the under-floor ducts from the other rooms. Then, natural convection delivers the heated air to the rest of the house. While the original idea was to use the duct system with an electric fan, the natural convection is quite good, so the fan has not been installed.

Some Construction Details

Waterproofing: The house's vertical surfaces have Volclay (bentonite) panels, and with bentonite driller's mud placed under the insulation on the curved ferrocement surfaces.

Insulation: Four 1-inch layers of extruded polystyrene (R-21) are installed down to a 7-foot depth. Three inches of insulation tapering to 1 inch are used the rest of the way down the walls. The exposed wall on the south side has 4½ inches of polyurethane foam (R-31) covered with a fiberglass-reinforced protective coating.

(continued on page 218)

Photo 8-2: This photograph of the Cooper house under construction shows the exterior walls of concrete block and the wire-mesh armature that will be sprayed with concrete to form the ferrocement roof.

Photo 8-3: Interior of the Cooper house.

Roof: There's a minimum of 4 feet of earth cover on the roof, with the depth increasing away from the centerline of the house. The strength of the curved ferrocement roof is sufficient to support a bulldozer during the backfilling. The roof reinforcement is ⅜-inch steel rebar laid in a grid pattern, wired together and covered with expanded wire mesh. This steel composite material is both the concrete form and the basic reinforcing for the curved roof component.

Walls: The first 5 feet are made of 8-inch concrete block grouted full and containing both horizontal and vertical steel reinforcing.

Wiring: Wiring is run in conduit where it is placed in concrete, although most of the wiring is placed in the wooden interior partitions using conventional wiring methods.

Plumbing: All supply pipe is copper. Most of the plumbing is inside cabinets or in wooden interior walls. Drain piping is PVC plastic.

Heating: South-facing windows provide direct-gain solar heat with a woodstove for backup.

Cooling: Short (10 feet) earth-covered cool tubes bring in outside air that is somewhat preconditioned by earth temperatures.

Ventilation: At the peak of the roof curve, a forced-air ventilator draws air either directly from the outside, through doors and windows or in through the cool tubes.

The house has been occupied since January of 1981, and the owners report that there are no problems. The energy performance is excellent with solar providing most of the heating. The back-up woodstove requires a single cord of wood for the heating season. Cooling is adequate for most summer days, although there were some days with excessive humidity in the house, which required dehumidification. There's no insulation under the floor slab, and where it was not covered by carpeting, some condensation occurred the first summer. Some dehumidification at strategic times cured the problem. Although building codes were not a problem, it was difficult for the owners to get a construction loan for the house, largely due to its unusual design. After construction was completed, permanent financing was obtained without added problems. The quietness of this house, its spacious quality despite its small size, and its economy of construction and operation make it a genuinely fine place in which to live.

Photo 8-4: The Cooper house is heated by direct-gain solar energy with a woodstove for backup.

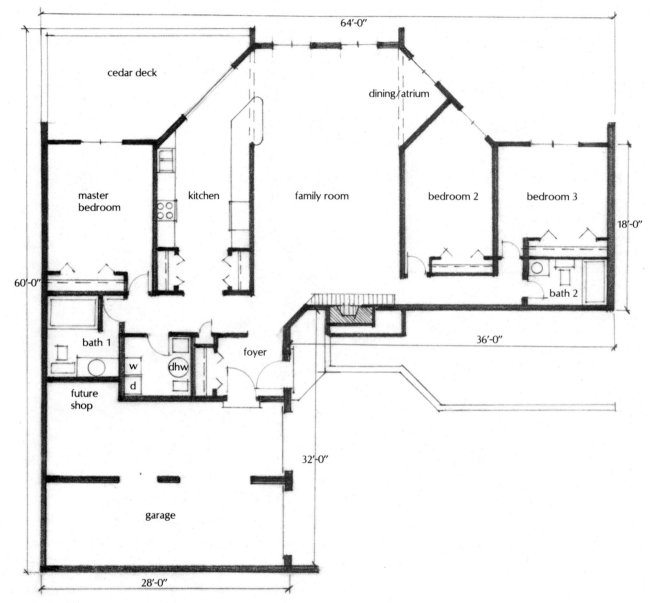

*Figure 8-4: The Freidrichs house
floor plan.*

The Freidrichs House

Wood construction is not common for underground houses, even though wood basements have been successfully built for decades. Everstrong Marketing, Redwood Falls, Minnesota, believes wood construction for underground houses is a very cost-effective approach. The home of Chuck and Mona Freidrichs is one of Everstrong's first treated-wood underground houses. This 3-bedroom, 1,900-square-foot house was enclosed in December, 1980, after about three months of construction, and is on a ¾-acre suburban lot in the north-central United States.

The CCA-treated wood shell was built by a contractor, although Chuck helped with the construction, and he and his family are doing most of the interior finish work. Although the Freidrichses had no problems obtaining a building permit, they could not secure a mortgage greater than 50 percent of the total construction cost. Getting even that much was difficult because their lending institution was not impressed by the merits of underground living. Insuring their house was easy, since Chuck works for an insurance firm, and Mona previously worked for one. Conventional stud and plate construction was used throughout, though the studs are unconventional 2 x 10s and 2 x 8s, and they are spaced on 12-inch centers (the studs in a conventional frame house are 2 x 4s or 2 x 6s and are usually 16 or 24 inches apart). The exterior surface is treated plywood that is glued and nailed onto the studs for strength and watertight integrity. The roof, which is under 18 inches of earth, uses laminated 2 x 12 beams. They are also spaced on 12-inch centers with every eighth one doubled for added strength.

Figure 8–5: Cutaway view of the Freidrichs house.

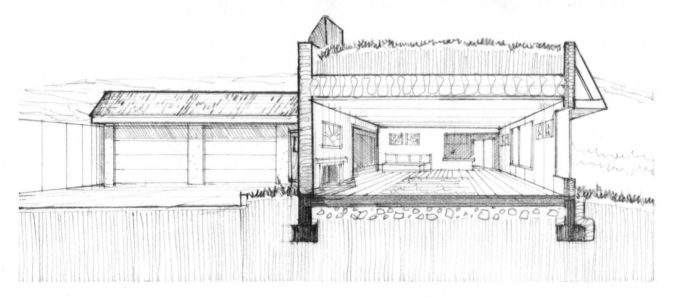

Photo 8-5 (above): *The southeast-
ern exposure of the Freidrichs
house.*

Photo 8-6 (right): *This view is the
"roof" of the Freidrichs house.*

The floor has 2 x 6 joists with ¾-inch plywood subflooring that is glued and nailed over them.

About a foot under the floor, there is a 6-inch layer of crushed rock that is insulated along its perimeter. The Everstrong design uses this underfloor cavity as a plenum for the gas heating system. The gas furnace dumps its heat into this plenum, and registers in the wood floor of the house allow the warm air to enter the living area. While this is not an unusual heat-distribution system for the Sun Belt, it is uncommon in cold climates because of the potentially large heat losses to the ground. Everstrong believes that this system works well in any underground environment if the plenum is well insulated. It certainly creates warm floors and eliminates the need for expensive ducts. Another advantage is that the same plenum can be used to distribute heat from a solar unit, although the Freidrichses' house does not have such a system.

Some Construction Details

Waterproofing: Bentonite is covered with a polyethylene plastic membrane.

Insulation: The underground walls have 10 inches of fiberglass, the ceiling has 12 inches, and the exposed wall has 6 inches of fiberglass plus 1 inch of polystyrene foam. There is a continuous polyethylene vapor barrier over the studs and rafters under the drywall.

Roof: Constructed of laminated 2 x 12 beams, the roof has 18 inches of earth cover.

Walls: The walls are of 2 x 10 stud construction on 12-inch centers, with treated plywood on the exterior.

Wiring: Conventional wiring systems are used throughout.

Plumbing: There are copper water lines and an ABS plastic drain system.

Heating: The furnace is natural gas, forced air. The house gets some solar heat from about 135 square feet of southwest windows.

Cooling and Ventilation: No special systems are installed.

In this 9,000 heating-degree-day climate, the largest monthly gas bill to date (which includes space heating, hot water and cooking) is $50, or about half that of neighbors with similarly sized, well-built, aboveground homes. This bill was for a month that had several nights colder than $-20°F$.

Photo 8-7: The interior of the Freidrichs house.

Would the Freidrichses change anything about their house if they could? Their only regret is that the gas furnace was installed with the capability of having a central air conditioner added later if needed. It isn't needed, and Mona wishes the extra equipment had not been installed, so that her laundry room could have been bigger. In general, the construction was straightforward, and any reasonably skilled do-it-yourselfer could build a similar house. The Freidrichses are advocates of underground wood houses and rate their house with an overall "excellent."

The Juenemann House

John and Henrietta Juenemann are very enthusiastic about their two-year-old earth-contact home. Although larger than average at 2,850 square feet, their house has energy bills more like a house one-third its size and, in the demanding climate of the Midwest, comfort levels better than any other house they and their four children have occupied. Certainly one reason they like it so much is that they took a very active part in its design and construction. Another is that they lavished a lot of attention on details, and more often than not, it is the details that make a good house into one that is outstanding. Also, the Juenemanns experienced no difficulties either in obtaining financing or in buying insurance for their home. In fact, when asked what he would do differently if he were to build his house again, John Juenemann responded, "Nothing. Absolutely nothing!"

The kitchen was carefully designed with an efficient layout and finely crafted cabinets. Each cabinet was built with extra storage units that pivot or slide into recesses that are usually wasted in kitchens. Another special feature is the provision for fire exits from the bedrooms on the north wall of the house, which is bermed along its entire length. Each of the bedrooms is connected to the one adjacent by a small doorway that passes through the closets. Thus, if people are prevented from leaving a bedroom by way of the hallway door, they can move through adjacent bedrooms via the emergency doors and bypass the fire danger. One of the more frequently used rooms in the Juenemann house is a large family/recreation room. It is there that the family rehearses for its musical performances as the Juenemann Family Singers. Adjoining this room are the open living, dining and kitchen areas. In these spaces, cathedral ceilings and strategically placed mirrors create a feeling of spaciousness.

Although it is hard to build an earth-contact house that is as energy efficient as a wholly underground house, the Juenemanns have done well. They thought out the possible problems in advance and took steps to prevent them. Unlike many earth-contact homes, the parts of the masonry walls that extend above the

Figure 8-6: The Juenemann house floor plan.

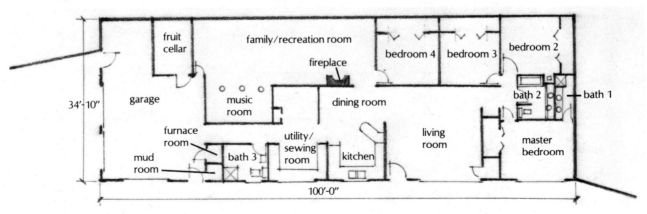

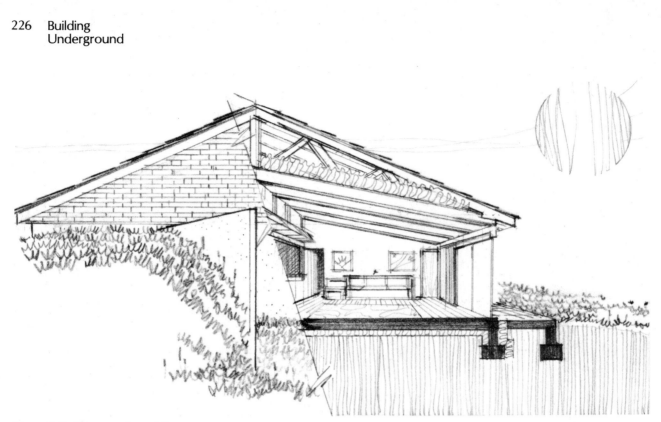

*Figure 8–7: Cutaway view of the
Juenemann house.*

ground are well insulated. This insulation is protected from rodents and mechanical damage by means of steel flashing that extends well below grade. Termites are warded off by the use of treated lumber at the junction of the concrete walls and the roof structure. The sometimes spotty coverage provided by batt-type insulation laid between rafters is overcome by a layer of blown-in insulation that covers the entire insulated ceiling area. Vents are installed in the frame gable end wall, as is the powerful whole-house fan, thereby eliminating extra penetrations through the roof and the possibility of leak problems. The fireplace heat exchanger is connected directly into the forced-air distribution system in the house, so the heat from the wood fire can be evenly distributed throughout the house.

Some Construction Details

Waterproofing: All masony walls have Bituthene self-adhering waterproofing membrane installed.

Insulation: The ceiling has 12 inches of fiberglass covered with 2 inches of blown-in rock wool. Walls that are exposed frame (gable ends and the south wall) have 6 inches of fiberglass. Masonry walls have 4 inches of Styrofoam insulation placed on the exterior down to 2 feet below grade.

Photo 8-8: The Juenemann house has a traditional roof and is bermed on three sides.

There are 2 feet of 2-inch foam below that, and the bottom 4 feet of wall have no insulation.

Roof: The traditional roof has heavy shake cedar shingles.

Walls: Masonry walls underground are 8-inch poured concrete with four horizontal and equally spaced ½-inch reinforcing bars and vertical bars every 4 feet. Wood walls are built with 2 x 6 studs and covered with brick veneer.

Wiring: Wires are strung in conduit where run in masonry, and conventional wiring systems are used elsewhere in the house.

Plumbing: All hot and cold supply pipes are copper placed inside flexible plastic pipe for protection where run under the floor slab. All drain lines are ABS plastic.

Heating: Although the house has a forced-air furnace that operates on propane, it is not used. Space heating is provided by solar energy and a metal-shell fireplace insert that feeds the distribution system installed with the furnace.

Cooling: Central air conditioning is installed but rarely used. Most cooling is provided by a whole-house fan, which is used at night to flush heat from the house and pull in cooler outside air.

Ventilation: Air movement is forced by the whole-house fan and a smaller kitchen vent.

Photo 8-9: The entire north wall of the Juenemann house is bermed.

Photo 8-10: Cathedral ceilings and south-facing windows create a sense of openness in the Juenemann house.

Photo 8–11: This south-facing window in the Juenemann house provides daylighting and direct-gain solar energy.

The house has proved to be inexpensive to operate. The central air conditioning has been used on the average for fewer than ten days a year when temperature/humidity levels have exceeded the cooling capacity of the whole-house fan. For back-up space heating, a propane-fired, forced-air furnace is installed, but the pilot light has never even been lit. The 190 square feet of south-facing windows provide direct-gain solar heat, and a metal insert-type fireplace meets the balance of the heat load. Despite the poor efficiency of the fireplace (when compared to an airtight woodstove), total yearly wood consumption in this 5,000-heating-degree-day climate is about 4 cords. While that is more than many of the homes I have visited, this house is also considerably larger than most, and with the poorer performance of the fireplace compared to a woodstove, I believe the actual energy consumption for the house is about that of a well-designed, fully underground house.

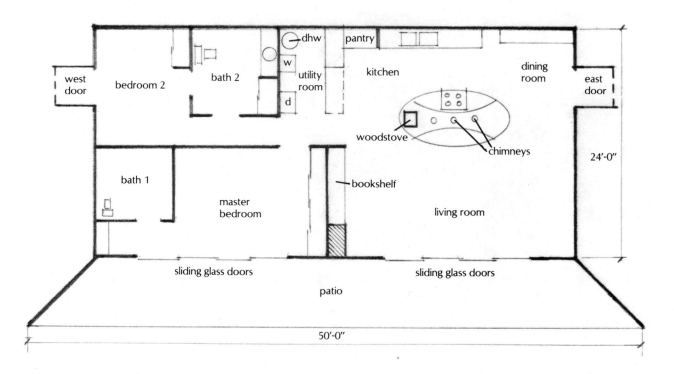

Figure 8–8: The Laughlin house floor plan.

The Laughlin House

In early 1980, Vernon and Jean Laughlin set out to find a house design that would combine comfort and energy efficiency for their central United States location, which has hot, humid summers and cold, windy winters. Their choice was an unusual design, developed by Terra-Dome Corporation, Independence, Missouri. Typically, Terra-Dome houses are made of 25-foot modules, which have vertical 8-foot sidewalls and an arched roof that creates a 12-foot-high ceiling at the center of the arch. Each module can be provided with a 16-foot archway for connecting adjacent modules, or it can be filled with windows or doorways to the outside. The secret to the system is in a special, patented forming methodology. Sidewalls are conventionally placed with aluminum forms. The arch, however, utilizes fiberglass forms that are mounted on a trailer with a hydraulic lift. The form is lifted into place and "locked in" for the pour; after curing, it is lowered and stored back on the trailer. The resulting barrel vault is engineered to support 8 feet of earth, plus a bulldozer! The wall and the vault are poured all at once, so there are no cold joints above the footing. The walls are 10 inches thick, and the vault is 14 inches at the base and 5 inches at the peak.

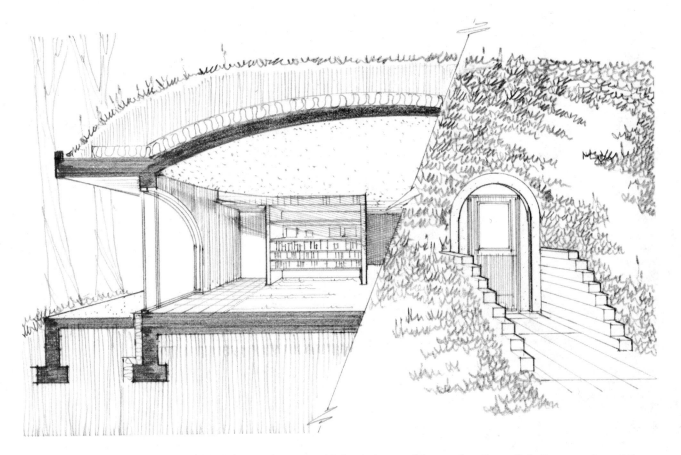

Figure 8-9: Cutaway view of the Laughlin house.

Continuous reinforcing is included in the package, and it is engineered for each installation.

Beginning in June, 1980, the Laughlins excavated and built their footings to Terra-Dome's specifications. As soon as the footings cured, the contractor placed the forms, poured the entire structure and removed the forms in a single week. The floor was poured later on by the Laughlins.

The Laughlins oriented their house to the south and instructed Terra-Dome to fill the exterior archways with glass for solar gain. Because the forming system leaves a smooth surface, the interior of the vault was covered with a light layer of textured material, thereby allowing the large concrete mass to store heat. Although some of the walls were covered, significant amounts of heat can still be stored in the structure.

Photo 8-12: The southern exposure of the Laughlin house.

Some Construction Details

Waterproofing: A liquid polyurethane elastomeric membrane was rolled on in two coats.

Insulation: Polyurethane foam was sprayed on to a thickness of about 2 inches. A plastic film covers buried foam surfaces, and a weatherproofing coat that looks like stucco covers areas exposed to the weather.

Roof and Walls: The 8-foot vertical sidewalls are 10 inches thick. The arched roof, which creates a 12-foot-high ceiling at the center of the arch, is 14 inches thick at the base and 5 inches thick at the peak. This system is designed to support over 8 feet of earth cover.

Wiring: Conduit is cast into walls and floors, and there are conventional runs in wooden interior partitions. Vernon, who works as a lineman for a large utility company, installed the wiring.

Photo 8–13: Terra-Dome modules create arched ceilings in the Laughlin house.

Photo 8–14: A woodstove provides backup heat in the Laughlin house.

Plumbing: All pipes are PVC or CPVC plastic placed in the concrete. Cold-water pipes are cast directly into the slab, while hot-water pipes are insulated. The drains are ABS plastic pipe. All plumbing is owner-installed.
Heating: A small, centrally located woodstove meets all heating requirements. An outside combustion-air source was not provided, but could have been.
Cooling: A small cooling load is met by a small, ½-ton (6,000 Btu/hr) air conditioner and by ceiling fans.
Ventilation: Air can enter or leave through doors on the south, west and east sides of the house.

The house's energy economy is everything the Laughlins expected. During the heating season, they use about 2 cords of wood, and in the cooling season, a small air conditioner keeps the house very comfortable. The primary comfort problem in the summer is humidity, but dehumidification is not a solution, since

Photo 8-15: This door provides egress to the main living areas from the east side of the Laughlin house.

the indoor air temperature is raised by a dehumidifier. The Laughlins found that a small air conditioner provides more than adequate comfort with no increase in electrical consumption over the dehumidifier.

A particularly notable feature of the house is its temperature stability. Even though the structure is heated only with a woodstove, during the record-low temperatures of January, 1982, the temperature in the house fell less than 5°F overnight.

In building the house, the Laughlins encountered no difficulty in getting a building permit or in obtaining a small mortgage. Since the mortgage company did not require insurance, the risk of damage to such a permanent, fireproof structure seemed so low that the Laughlins elected not to insure at all. Perhaps the best testimony is from Vernon Laughlin, who said, "The house performs just like they [Terra-Dome] said it would. I admit I am surprised and very pleased."

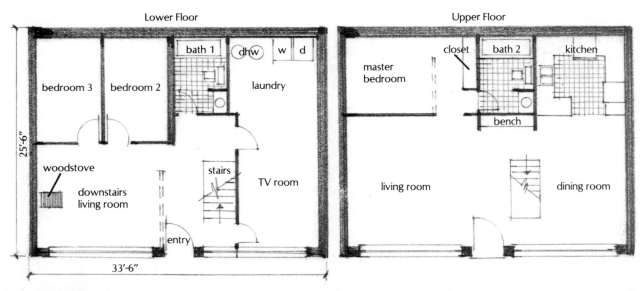

Figure 8-10: The Miley house floor
plan.

The Miley House

Although two-story underground houses are uncommon, James and Nina Miley like the advantages offered by their design. Their house was designed by Nina and an architectural student and friend, Will Simmons, to be as close to zero energy consumption as possible at their 4,600-heating-degree-day Great Plains site. By building a two-story house, the Mileys were able to get direct sunlight into each room of the house without the long horizontal sprawl necessary for the same result with a single story. Also, the convective air circulation patterns created by the two stories allow them to heat the 1,500 square feet of living space with a small woodstove. Likewise, in the summer, rising hot air is easily channeled through the two monitor-style skylights.

To the Mileys, the proof of a successful design is their use of less than a cord of wood through the winter, and maximum summer temperatures no higher than 80°F even when it's over 100°F outside — that's with neither dehumidification nor air conditioning. For their climate, which has cold, windy winters and hot summers, that is good performance indeed. Much of the credit for the winter heating has to go to the 180 square feet of south windows for direct-gain heating. The windows are covered at night with Nina's home-sewn insulating curtains. Summer operation also benefits from the curtains by limiting heat gain. Recorded relative-humidity levels for summer have been in the comfortable range of 40 to 50 percent.

The high thermal stability of the house is due primarily to the large areas of exposed concrete. The cast-in-place walls were formed carefully, and seams were easily removed, leaving a surface that, when painted, actually looks like

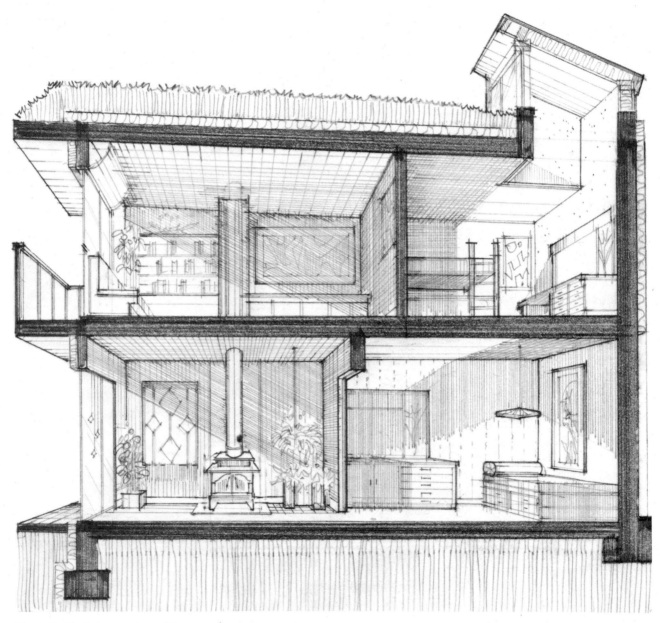

*Figure 8-11. Cutaway view of the
Miley house.*

quality plaster. The roof was made with 10-inch, hollow core, precast, prestressed panels with 2 inches of additional concrete poured over the top to aid in waterproofing and strength and to allow reinforcing ties to the vertical walls.

Photo 8–16: The southern exposure of the two-story Miley house.

Some Construction Details

Waterproofing: An EPDM membrane is continuously bonded to the concrete. The contractor also used a basement damp-proofing compound before applying the butyl membrane.

Insulation: There are 3 inches of expanded polystyrene on the roof and 1½ inches on the exposed south wall (furred out with 2 x 2s). There are 2 inches on the second-story walls and none on the lower-story walls.

Roof: The roof drops about 1 foot for each 12 feet of horizontal run. The earth cover is about 1 foot at the south edge and increases to 2 feet at the north wall of the house.

(continued on page 239)

Photo 8-17: The monitor skylights on the Miley house provide day-lighting and ventilation.

Photo 8-18: The lower floor of the Miley house.

Photo 8-19: The upper floor of the Miley house.

Walls: The walls are poured concrete with steel reinforcing engineered for the site. All concrete work was contracted, although some labor was provided by the family.

Wiring: Wiring is run in wood interior walls, in conduit cast into the walls and floors, and through the hollow cells in the precast ceiling panels.

Plumbing: Copper water pipes run in a single plumbing wall. Drains are ABS plastic running to a private septic system.

Heating: Solar energy provides most of the heat, with a woodstove for backup.

Cooling and Ventilation: Natural cooling and ventilation are used in the house.

The monitor skylights are particularly valued by the Mileys. They are located at the north wall of the house and provide nicely diffused illumination and excellent ventilation for the kitchen and the master bedroom. During the winter, the Mileys install an additional plastic layer at the ceiling opening to the monitors, which reduces heat loss without much reduction in available light or solar heat gain.

Photo 8–20: The south-facing windows of the Miley house allow direct-gain solar energy and daylighting.

There were no problems with building permits nor with financing, one reason being that the Mileys had previously built a pioneering passive solar home and had established a track record with their lending company. The Mileys had no problem in insuring the structure, and the cost of their insurance is the standard rate.

What would be different if the Mileys were to build another earth-sheltered house? They would add a small circulation fan to even out the temperature difference between the two floors—although it is now only about 5°F. Also, James said he would hesitate to build a speculative two-story house, since the extra excavation and the high concrete walls add difficulty and could be prohibitively expensive.

The Miley family did all of the interior work and much of the exterior work, except for the concrete and placement of the massive stone, terraced retaining walls. The total construction cost was about $50,000 in 1979.

The Mileys advise other home-builders "to have a clear idea of what you are going to do before starting. Get competent help in those construction trades that you are unfamiliar with. Don't just wing it."

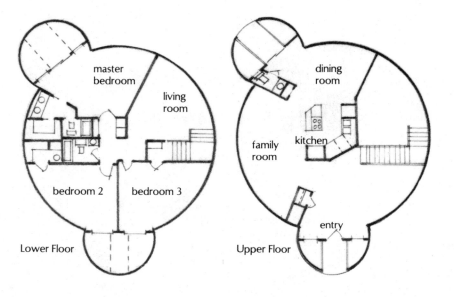

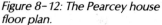

Figure 8-12: The Pearcey house floor plan.

Source: Redrawn from plans and
designs under copyright © with
the permission of Earth Systems,
Phoenix, Ariz.

The Pearcey House

In the hot desert country of the Southwest, many Navahos live in hogans, which are basically domes made from crisscrossed logs. In many cases, these hogans are covered with earth to help ward off the heat of the day and the chill of night. Brothers Dale and Gene Pearcey have done much the same thing in designing a house for Dale and his wife, Gail, except that it is made of ferrocement instead of logs. Since much of the cost of an underground house is in the massive concrete structure, the brothers reasoned that a dome shape would be strong enough with much thinner concrete than that required for the more conventional rectangular shape. Therefore it could be built for less money. Their ferrocement house was such a success that the Pearcey brothers founded Earth Systems, Inc., Phoenix, Arizona, to market the hemispherical, underground-house building system they developed.

The construction of the Pearcey house began with the placement of ten curved steel ribs. The ribs were attached at the upper ends to a 10-foot steel ring and at their bases to a concrete footing. A larger steel ring was added about two-thirds of the way down the ribs to support the upper floor, and then a mat consisting of a layer of fabric (much like burlap), reinforcing bars and steel mesh was placed over the ribs. An additional "pillbox" was built onto the 10-foot steel ring, forming a monitor section that ultimately would be above grade. The concrete was then sprayed on to a thickness of 4 inches and allowed to cure.

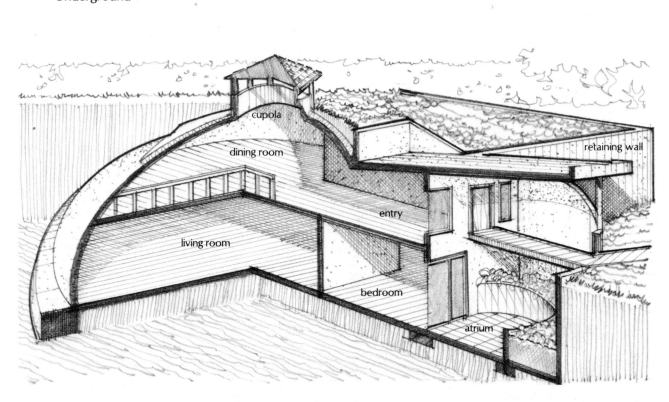

Then the shell was waterproofed, insulated and backfilled. Since no forms were used, there was no form expense. Also, since the shell is less than half the thickness of conventional, rectangular house shells and because the hemisphere encloses the most volume for the least surface area of any shape, the basic cost of this shell was considerably lower than that of comparably sized rectangular shells.

Another advantage the Pearceys attribute to their design is ease of cooling. The monitor at the peak of the dome collects rising hot air and exhausts it to the outside, while drawing in cooler, ground-level air. For summer operation, the monitor is opened at night to draw in cooler night air. The monitor is closed off during the heat of the day, then reopened at night. Some air conditioning is used in the house, but cooling costs are a third or less than that of an aboveground house of comparable size. Exterior foam insulation was placed only at the topmost part of the dome. As you know from reading this book, the heat of the sun penetrates 10 or more feet into the desert soil, so houses that the Pearceys build in the future will have insulation down to at least the 10-foot level. This should further reduce the cooling load.

Two more advantages of the Pearcey's design are good daylighting and ease of construction for do-it-yourselfers. The centrally located monitor and the sloping, white interior dome surface provide unusually even illumination through-

Photo 8-21: The Pearcey house.

out the house. Do-it-yourselfers find the Pearceys' design attractive because there is no elaborate forming system, so preparing for concrete application is much simpler than for other types of concrete houses. Also, with this house design, the day the concrete is applied, the shell is completed. There are no multiple pours spread over days or even weeks, and there are no delays in waiting for the concrete to cure before the forms are removed, because there are no forms. Most do-it-yourselfers will probably have to hire contractors to spray on the concrete, however, since proper application requires considerable experience.

Convincing building officials and lending institutions that building an underground house is a good idea is a problem by itself without the additional problem of an unusual design. But these were not serious problems for the Pearceys. Their house was built in an area that has strict building codes, and it was approved by the Federal Housing Administration (FHA) for financing. But others building a similar home may have to persist to overcome the natural inertia of some zoning boards and lending establishments.

Photo 8-22: The centrally located monitor skylight on the Pearcey house provides daylighting throughout the house.

Some Construction Details

Waterproofing: Bentonize is covered with a polyethylene sheet.

Insulation: There are 2 inches of extruded polystyrene about 3 feet down from the peak of the dome. None is installed below that.

Roof and Walls: There are 4 inches of concrete sprayed over a layer of fabric, reinforcing bars and steel mesh placed over ten curved steel ribs. The amount of earth cover varies but is about 3 feet at the top of the dome.

Wiring: Wires are in conduit only where run in the concrete.

Plumbing: Copper water pipes and ABS plastic drains are used.

Heating: A heat pump is installed but little used in this climate.

Cooling: Natural ventilation is used when possible, but a heat pump is available for extreme conditions.

Ventilation: Doors at the lower-floor level and operable windows in the monitor provide good ventilation when the temperature of the house is higher than that of outdoors.

Although Dale and Gail Pearcey cannot be considered unbiased in their praise for the house, since they are in business to sell house kits of the same design, their unbiased electric bills show that their energy consumption is one-third that of a comparably sized, conventional aboveground house. Also, there is no doubt that their claims of good daylight and spacious interiors are true. So, I believe Dale and Gail, biased or not, when they say, "After 2½ years in this house, it is by far the best we have ever lived in."

Photo 8-23: The skylight provides even lighting in the Pearcey house.

Photo 8-24: The interior surfaces of the dome are white to allow natural light to illuminate the interior of the Pearcey house.

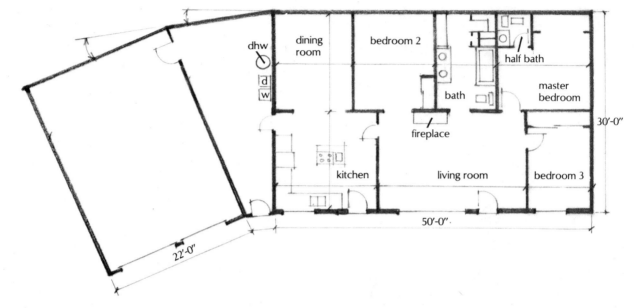

Figure 8–14: The Rogers house floor plan.

The Rogers House

Representing the classic poured-concrete house is the home of Greg and Belva Rogers. The house, occupied in the spring of 1979, was built in one year by Belva and Greg (who was then working for a foundation construction company), Greg's father, Duane Rogers (an ironworker), and Belva's parents, Vernon (a pipefitter) and Lavoy Hart. Located in the 4,500-heating-degree-day climate of the central Midwest, the house was designed for efficiency in both heating and cooling.

Construction costs added up to $37,500. This is for a 2,460-square-foot design, of which about 1,500 square feet are heated living space, with the rest devoted to storage space and a garage. Of course, there is little labor cost in that figure, since the only contracted work was for drywall finishing and laying some decorative brick on the fireplace and on the exposed south wall.

As the floor plan shows, this home has no space lost in hallways. The Rogerses designed their house without halls both to maximize the useful living area and to ensure that light from the south-facing windows could reach all rooms.

Although there is no part of the structure specifically designed for solar heating, its orientation is about 15 degrees west of true south, and the south-facing windows are appropriately sized for good direct-gain heating. While the concrete ceiling provides good thermal storage, most of the concrete walls are covered with drywall over strips of ¾-inch plywood (left over from the forms) and are less effective as thermal flywheels than they could be if the concrete were left exposed.

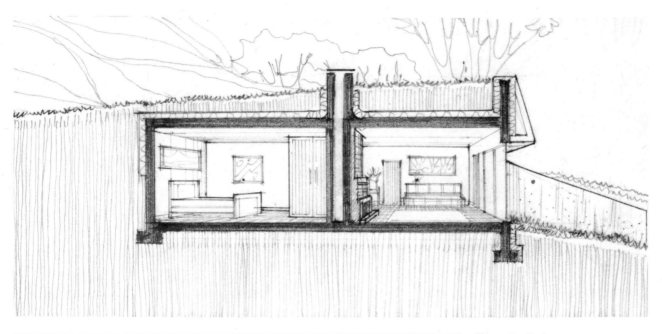

Figure 8-15: Cutaway view of the Rogers house.

Some Construction Details

Waterproofing: A membrane composed of polyethylene with rubberized asphalt on one side is bonded to the concrete exterior under the insulation.

Insulation: There are 2 inches of expanded polystyrene on the roof (which also has 3½ feet of earth cover) and 2 inches extending down the walls halfway to the footings. On the southern side, footings are also insulated.

Roof: The roof is pitched with a total drop of 2 inches from south to north. The steel reinforcing consists of two layers of crisscrossed ½-inch rebar.

Walls: Poured concrete is 8 inches thick, with ½-inch rebar 4 feet on center vertically and 3 feet on center horizontally. There's additional reinforcing over doors and windows for extra strength.

Wiring: Wiring is mostly UF (underground-feeder) grade cast directly into the floor slab without conduit. There are some conventional wiring runs in the wooden interior walls.

Plumbing: All cold-water piping is PVC plastic, hot water is CPVC plastic, and drains are ABS plastic. Piping is cast into the floor slab with no problems reported so far. The sewage treatment facility is a private lagoon 300 feet from the house.

Heating: Baseboard electric units are installed under each of the windows.

(continued on next page)

Photo 8–25: Light entering the south-facing windows of the Rogers house can reach each room.

A centrally located masonry fireplace provides a large amount of heated thermal mass.

Ventilation: Besides operable windows, there are registers in the central chimney to exhaust heated air when necessary. By running this and all plumbing and kitchen vents to the central stack, only one roof penetration is needed for the entire house.

The performance of the house has lived up to the Rogerses' hopes, with $17 worth of electricity being their highest monthly heating bill thus far. Cooling is handled with a single window-type air conditioner on those few days when natural cooling is inadequate. Although interior summer humidity levels have fallen since the summer of 1979 when the concrete was still curing, a few hours of dehumidification each week in the hottest months of the summer are needed for optimum comfort.

The location of the house in a rural area that has no building codes eliminated any problems in dealing with out-of-date building codes, but no local lending institutions were willing to provide a mortgage, so the Rogerses relied on private financing. Also, in 1978 insurance companies were not generally aware of the advantages of underground construction, and not only was it difficult to get homeowner's coverage, but the Rogerses also had to pay higher-than-standard rates. When Greg asked the agent why a fireproof, windproof house cost more to insure than a similarly sized wood-frame, aboveground house, he said that a ''B-52 might crash into it or something.'' Since the nearest large airport is 35 miles

Photo 8-26: The interior of the Rogers house.

Photo 8-27: The fireplace in the Rogers house has two roles: to produce heat and to store heat in its thermal mass.

away, that seemed a bit unlikely to the Rogerses. Fortunately, the 1980s finds considerably more enlightenment with both lenders and insurers.

All in all, the Rogerses' house works very well, and when asked what they would do differently, their answer was a simple "Nothing." Greg did say, however, that being an owner-builder is not easy, and his best advice is, "Don't give up. It's worth it."

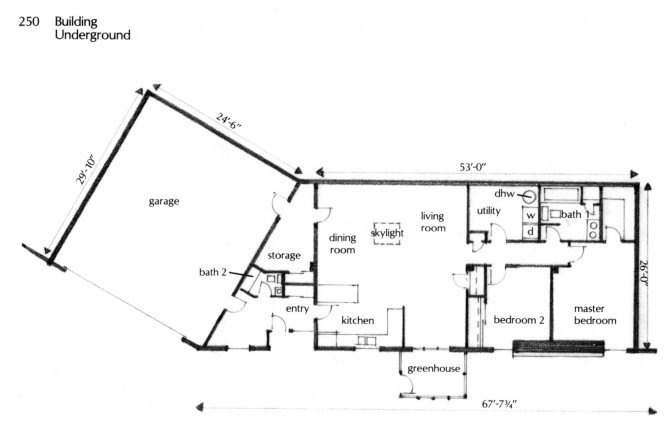

Figure 8–16: The Shepard house floor plan.

The Shepard House

The prestressed, posttensioned concrete residential designs of Lon Simmons of Simmons & Sun, Inc., High Ridge, Missouri, have found their way into every geographical area of the United States. Bob and Lucille Shepard, faced with the high energy costs and severe winter weather of New England, selected one of Simmons' solar designs because they wanted a house with maximum thermal comfort at reasonable cost.

Their house incorporates all three classes of passive solar components. The bedrooms have both south-facing windows and Trombe walls. The open living-kitchen-dining area has direct-gain windows and a sunspace. These systems are the primary heat sources, while the centrally located woodstove provides auxiliary heat. The heat-storage capacity of the concrete is increased by the use of an interior stucco finish on the walls, which is more effective in absorbing solar radiation than standard drywall. The ceiling, which includes a 26-foot unsupported span (made practical by the prestressing process), is finished with a spray-on textured material over a thin coat of drywall compound. This texturing was done to even out any roughness in the concrete resulting from slightly misaligned form joints.

By using prestressed, posttensioned concrete, wall thicknesses of 8 inches and a roof thickness of 10 inches were practical despite the long open spans and

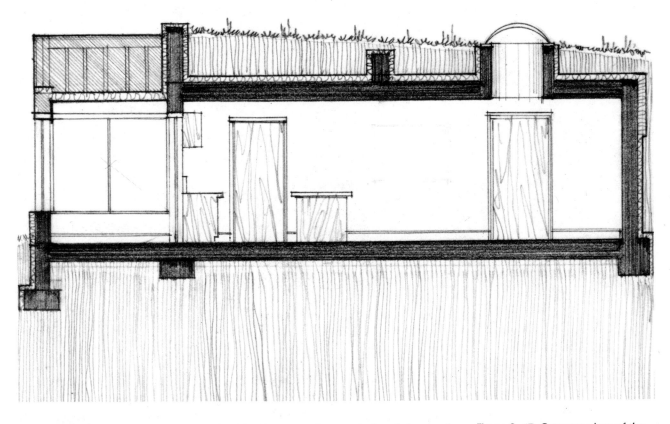

Figure 8–17: Cutaway view of the Shepard house.

the 3 feet of soil on the roof. To further increase the strength of the roof, a 2-foot-high, 16-inch-wide concrete beam is included at the roof center.

South-facing windows and a 3-foot skylight that is mounted in the living room roof provides plenty of daylighting for most activities. The skylight well is a 3-foot, precast-concrete pipe that was placed on the roof form before the concrete pour.

The sunspace acts both as a solar heating system and as an airlock entryway. Glass doors between the sunspace and the living area provide light as well as a view from the living room.

The Shepard house, which was built in 1981, has 2,400 square feet of total area with about 1,600 square feet of living space. During their first winter, the Shepards found they could easily maintain comfortable conditions even when ambient temperatures were −5°F. These proud owners call their home "the most affordable place we have ever lived in." Summer operation presents no problems because their summer weather is quite moderate, and the solar components have overhangs to provide shade.

Photo 8–28: The Shepard house.

Photo 8–29: The south-facing windows and the skylight provide daylighting in the Shepard house.

Photo 8–30: The Shepard house has three passive solar features: south-facing windows, a sunspace and a Trombe wall.

Some Construction Details

Waterproofing: Bentonize is on the walls and bentonite on the roof. Waterstops are the Waterstop-Plus brand, which is also a bentonite product.

Insulation: The roof has 6 inches of expanded polystyrene. The top 4 feet of the wall are insulated with 6 inches of expanded polystyrene, while the lower half has 2 inches. There is no insulation under the floor, but a continuous vapor barrier separates the floor from the ground. The exposed south wall is furred out with 2 x 6 studs on the outside and has 6 inches of fiberglass insulation. The south footings are insulated with 2 inches of expanded polystyrene.

Roof: The concrete roof is 10 inches thick and has 3 feet of earth cover.

Walls: The concrete walls are 8 inches thick.

Wiring: Wires are in conduit where they pass through the concrete. Conventional wiring methods are used in the wood walls.

Plumbing: Copper pipe under the floor is used for all hot and cold water pipes. All drains are ABS plastic.

Heating: The multiple passive solar components provide most of the heating, with a woodstove for backup.

Cooling and Ventilation: Natural ventilation gets a boost from kitchen and bath vent fans.

Photo 8-31: Interior of the sunspace in the Shepard house.

Photo 8-32: Movable insulation on the windows in the Shepard house keeps heat in at night. The inlet and outlet vents are for the Trombe wall.

Bob Shepard, a concrete contractor, and his crew did the construction. Since this was his first posttensioned concrete structure, it took more time to complete than it normally should. But despite the increased cost of labor, the house was built for essentially the same per-square-foot rate as a comparably sized aboveground house. Would he do it again? He answered, "Yes, sir! This is the first posttensioned house in New England, and I'm looking forward to building many more."

Estimating Heating Requirements for Underground Houses

The estimation of heat loss from an aboveground frame house is not always a simple matter, but at least the conduction of heat through the skin of the structure is readily calculable. The underground house presents a much more complex problem. Part of the added difficulty stems from the change in soil temperature with depth. On the surface, any heat lost from the skin of a house is transferred away so quickly that the entire surface of the house can be considered to be at the temperature of the outside air. Underground, the temperature of the earth varies with depth in a complex fashion, with all sorts of time-delay effects confusing the situation. Then, too, the rate at which heat moves through the earth is slow enough that the mere presence of a body (a house) at a different temperature changes the soil temperature. That change is then acted on by the surrounding soil, which is at a still different temperature. Clearly, this is a problem for a clear-thinking physicist armed with a fast digital computer. Indeed, the only practical way to get a handle on the complex heat-flow patterns and temperature differentials around an underground structure seems to be with elaborate computer simulations. Yet, even those leave a lot to be desired when the structure is not a simple, symmetrical shape. I have faith that within a few years remarkable strides in personal computer systems will occur, but it seems unlikely that computer simulations of heat-flow calculations will be of value to the man in the street for quite a while.

Of course, part of the reason that the computer simulation method is so difficult is that the machine is determining the heat-flow rate and how that heat flow changes the surrounding conditions. Then it's recomputing a new heat flow to fit the changed conditions — which further changes the conditions — and so on into a dizzying spiral of approximations. This level of complexity is necessary if the heat-flow rates are to be computed for an entire season, but for the simpler purpose of figuring out how big an auxiliary heating system must be to keep the house at comfortable temperatures under the worst-case conditions, only a "snapshot" of the surrounding soil conditions needs to be used rather than the "movie" needed for the long-term determination. If we had a snapshot of the worst-case (lowest average temperature) soil condition, then the calculation would be essentially the same as that used for the aboveground house, with the minor complication that an allowance would have to be made for the change of temperature with depth.

Unfortunately, the snapshot method presupposes that you know what the worst conditions will be. It is not very likely that you do. Usually, the only way to know is to monitor the soil temperatures at many depths and then feed that data to a computer, which will compensate for the presence of the house. Perhaps in the future, enough data will be available to allow the use of standard temperature/depth curves for various sites, but it isn't possible yet. Frankly, I have given up trying to calculate heat losses — even in aboveground houses. The basic reason is that for very energy-efficient houses, factors that have little to do with outside temperatures have so great an effect that errors in static heat loss calculations of 50 percent and more may be unimportant. In underground houses, the energy requirement to make up for conduction of heat to the surrounding earth is so small that the presence or absence of a frost-free refrigerator will change auxiliary heating needs significantly. On a cold, windy day, the presence or absence of a vestibule entry may make more difference in heating needs than an extra 6 inches of insulation. Even with aboveground houses, whose heating loads are clearly dominated by the losses through the skin of the structure, family life-style has a significant effect on energy efficiency. With high-performance housing, the variations in the way people live are little different from those found in poorly performing houses, but the variations will show up as a much larger percentage of the total heat load in the energy-efficient houses.

As everyone knows, when science fails — guess. When the method for guessing has been codified and somewhat validated, it becomes a rule of thumb. The rule of thumb I use for estimating heating needs for an underground house is based on data provided by occupants of many different types of houses and uses the concept of Home Heating Index (HHI). Figure A–1 graphically presents this rule of thumb. The assumption is that the energy usage of underground houses will fall into a standard statistical distribution curve. My observation is that for underground houses in use today, the peak of the distribution curve falls at about 6 Btu per square foot per heating degree-day (6 Btu/ft^2/HDD). That means that for every heating degree-day, every square foot of heated space in the house will need 6 Btu to keep it at comfortable temperatures. Some underground houses need as much as 10 Btu/ft^2/HDD for the same performance. They are the ones with lots of nonsolar glass and poor insulation, or they belong to families with life-styles that aren't conducive to conservation. At the other end is the house that has only south-facing glass and is not only well insulated but is occupied by a very energy-conscious family. It may need less than 1 Btu/ft^2/HDD to maintain comfortable temperatures.

The following rule-of-thumb estimator can be used two ways. It can provide you with an estimate of the probable heating requirements for the house on a specific design-day, and it can be used in conjunction with the graph in figure 3–8 for sizing direct-gain systems for top economic performance.

To guess at the size of the heating system needed, pick a spot on the graph that seems to describe your situation and choose an HHI number from the chart. Then, choose a design-day temperature — one that you would consider to be the average temperature of the coldest two or three days of a typical year. Subtract

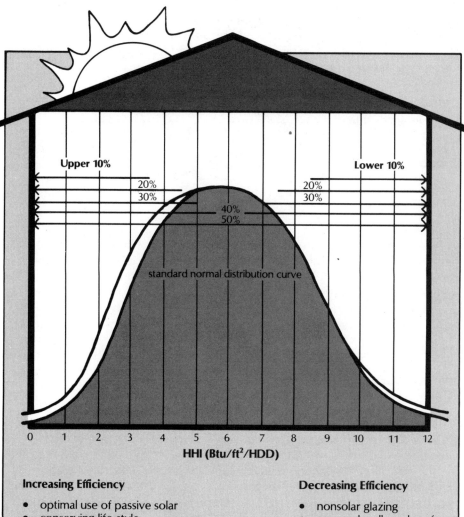

Increasing Efficiency

- optimal use of passive solar
- conserving life-style
- quality construction
- no nonsolar glazing
- decreasing exposed surfaces
- exterior insulation
- use of economizer
- exterior movable insulation (night)

Decreasing Efficiency

- nonsolar glazing
- exposed walls and roof
- poor construction quality
- nonconserving life-style
- inadequate insulation
- no movable insulation

Figure A-1: Graphical determination of Home Heating Index (HHI) using the ROT (Rule of Thumb) method.

that design temperature from 65 to get the heating degree-days for that cold day; multiply that by the HHI you've chosen from the chart, and then multiply the result by the number of square feet of your house that are heated. The end result is the estimated total Btu's needed to keep the house comfortable on that design-day. If you divide that by 24, you will have the Btu's needed per hour (Btu/hr), which is the way most furnaces are rated. Many people add about 30 to 50 percent to that figure as an insurance factor and to compensate for lowered efficiency as the years go by. The formula looks like this:

$$\text{Design heat load} = \frac{(65 - \text{design temperature}) \times (\text{Btu/ft}^2/\text{HDD}) \times (\text{square feet of heated floor area})}{24}$$

or

$$\text{Btu/hr} = \frac{\text{HDD} \times \text{HHI} \times \text{ft}^2}{24}$$

As an example, assume you have a 1,500-square-foot house in a climate that has an average temperature on its coldest days of about 10°F. The HDD for the design temperature is 65 − 10 = 55 degree-days. The house is all underground except for the glazed south wall, which is also well insulated. You have a conservation-conscious family. You look on the chart and choose an HHI of 3, which is above average but not extreme. Using the formula above, the example looks like this:

$$\frac{(65 - 10) \times 3 \times 1,500}{24} = \frac{247,500 \text{ Btu/day}}{24} = 10,000 \text{ Btu/hr}$$

That is an amount easily provided by a small woodstove or even a standard water heater — but it is realistic for a good house occupied by energy-conscious people. I have divided the normal distribution curve into percentile ranges to aid in your estimation of your house. If you think you have a house and family in the top 10 percent in energy efficiency, then choose an HHI of 3.5. If you are running with the pack, then choose an HHI of 6. If you have opted for a lot of nonsolar glass and you have ten kids constantly running in and out of the house, put yourself in the lowest 10 percent, with an HHI of 8.75.

How good is this method? I feel it's about as good as any other that is available to the layperson. Hopefully, in a few years enough data will be available to allow creation of simplified versions of the current computer models that can be used with a hand calculator or small personal computer and still yield good results.

Manufacturers of Waterproofing Materials

	TRADE NAME	MANUFACTURER
Cementitious Products	Anchor Masonry Surfacer	Anti Hydro Co. 265 Badger Ave. Newark, NJ 07108 (201) 242-8000
	Foundation Coat	Conproco Corp. P.O. Box 368 Hooksett, NH 03106 (603) 668-8810
	Hey'Di K-11	Hey'Di American Corp. 2801 Crusader Circle Virginia Beach, VA 23456 (804) 468-2200
	Thoroseal	Thoro System Products 7800 Northwest 38th St. Miami, FL 33166 (800) 428-6652
Liquid-Applied Membranes	A H Seamless Membrane	Anti Hydro Co. 265 Badger Ave. Newark, NJ 07108 (201) 242-8000
	Duraflex M and Duralastic	Dural International Corp. 95 Brook Ave. Deer Park, NY 11729 (516) 586-1655
	Duramen	Pecora Corp. 165 Wambold Rd. Harleysville, PA 19438 (215) 723-6051

	TRADE NAME	MANUFACTURER
Liquid-Applied Membranes *(continued)*	Extra Tite	Gemstar Building Materials Co. P.O. Box 800 Dallas, TX 75221 (214) 659-9800
	Gacoflex	Gaco Western, Inc. P.O. Box 88698 Seattle, WA 98188 (206) 575-0450
	Geocel Exterior Brushable Sealant and Geocel Water Shield	Geocel Limited, Inc. P.O. Box 398 Elkhart, IN 46515 (219) 264-0645
	HLM 1,000 Series and HLM 3,000 Series	Sonneborn Building Products Contech, Inc. 7711 Computer Ave. Minneapolis, MN 55435 (612) 835-3434
	Hypalon	E. I. du Pont de Nemours & Co., Inc. 10th and Market Sts. Wilmington, DE 19898 (302) 774-1000
	Liquiseal	Carlisle Tire & Rubber Co. Division of Carlisle Corp. P.O. Box 99 Carlisle, PA 17013 (717) 249-1000
	One-Kote	Karnak Chemical Corp. 330 Central Ave. Clark, NJ 07066 (201) 388-0300
	Raylite	Raylite Co. P.O. Box 7218 Wilmington, DE 19803 (302) 652-3904

TRADE NAME	**MANUFACTURER**
Tremproof 50 and Tremproof 60	Tremco, Inc. 10701 Shaker Blvd. Cleveland, OH 44104 (216) 229-3000
Vulkem	MAMECO International 4475 E. 175th St. Cleveland, OH 44128 (216) 752-4400
Wall Up	Conklin Co., Inc. 4660 W. 77th St. Edina, MN 55435 (612) 831-4044

Montmorillonite Clay (bentonite)

Bentonize R–80–S, Bentonize R–80–T, and Waterstop-Plus	Effective Building Products, Inc. 24100 Chagrin Blvd. Suite 470 Beachwood, OH 44122 (216) 831-7044
Volclay	American Colloid Co. Building Materials Dept. 5100 Suffield Ct. Skokie, IL 60077 (312) 966-5720

Preformed Sheet Membranes

Hypalon	E. I. du Pont de Nemours & Co., Inc. 10th and Market Sts. Wilmington, DE 19898 (302) 774-1000
Sure-Seal	Carlisle Tire & Rubber Co. Division of Carlisle Corp. P.O. Box 99 Carlisle, PA 17013 (717) 249-1000
Vaporseal	The Noble Co. 614 Monroe St. Grand Haven, MI 49417 (616) 842-7844

	TRADE NAME	MANUFACTURER
Roll Goods	Bituthene	W. R. Grace & Co. 62 Whittemore Ave. Cambridge, MA 02140 (617) 876-1400
	Laurenco Waterproofing Systems and Products	Laurent Corp. P.O. Box 244 Twinsburg, OH 44087 (216) 425-7888
	Poly-Mat	Karnak Chemical Corp. 330 Central Ave. Clark, NJ 07066 (201) 388-0300
	Premolded Membrane and Melnar	W. R. Meadows, Inc. P.O. Box 543 Elgin, IL 60120 (312) 683-4500
	Yellow Jacket	Gemstar Building Materials Co. P.O. Box 800 Dallas, TX 75221 (214) 659-9800

SOURCE: James J. Brown and Terry A. Johnson, *Earth Shelter Waterproofing: The Guide to Total Moisture Control* (Portsmouth, Ohio: Underground Homes, 1980).

Abercrombie, Stanley. *Ferrocement: Building with Cement, Sand, and Wire Mesh.* New York: Schocken Books, 1977.

Allen, Edward. *Stone Shelters.* Cambridge, Mass.: MIT Press, 1969.

Allen, Jonathan M. "A 100% Passive Solar Underground Greenhouse," *Alternative Sources of Energy,* September/October 1980, pp. 8–13.

American Concrete Institute. *ACI Manual of Concrete Practice: Recommended Practice for Measuring, Mixing, Transporting, and Placing Concrete.* Detroit, Mich.: American Concrete Institute, 1978. Order no. ACI 304–73.

American Plywood Association. *APA Design/Construction Guide: All-Weather Wood Foundation.* Tacoma, Wash.: American Plywood Association, 1978.

American Society of Civil Engineers (ASCE). *The Use of Underground Space to Achieve National Goals.* New York: ASCE, 1972.

American Society of Heating, Refrigerating and Air-Conditioning Engineers (ASHRAE). *ASHRAE Handbook of Fundamentals.* Atlanta, Ga.: ASHRAE, 1981.

American Underground Space Association; Minnesota Society American Institute of Architects; and Minnesota Society of Professional Engineers, cosponsors. *Collected Papers Presented at Earth Sheltered Housing Conference and Exhibition,* Minneapolis, Minn., April 1980. Available from Underground Space Center, University of Minnesota.[1]

Anderson, Brent. "Investigating Soils." *Architecture Minnesota* 6 (April 1980):39–41.

Anderson, Carol. "Earth-Sheltered Housing from the Ground Up." *Builder,* 1 August 1980, *pp. 44–59.*

"Back to the Earth." *House and Garden,* April 1980, pp. 158–61.

Barker, Michael. *Building Underground for People: Eleven Selected Projects in the United States.* Washington, D.C.: American Institute of Architects, 1978.

Barnard, John E., Jr. "The Cape Cod Cottage and the Earth Sheltered/ Passive Solar Home." *Architecture Minnesota* 6 (1980):49–50.

Baum, Gregory T.; Boer, Andrew J.; and Macintosh, James C., Jr. *The Earth Shelter Handbook.* Milwaukee, Wis.: Tech/Data Publications, 1980.

For Further Reading

Prepared by
LaVerne Koelsch Jones
Librarian
Architectural Extension
Oklahoma State University

Beadles, Joyce. "The Earth Has Won Our Hearts." *Mother Earth News,* July/August 1978, pp. 96-99.

"The Beale Solar-Heated Subterranean Guest House." *Mother Earth News,* May/June 1977, pp. 80-81.

Bell, Jackie Lee. "Consumer Attitudes toward an Earth-Insulated Solar House and a Solar Greenhouse Residence." Master's thesis, Oklahoma State University, 1979. Available through interlibrary loan.

Bethany, Marilyn. "Earthy and Efficient." *New York Times Magazine, 21 October 1979, p. 114.*

Bice, Thomas N. "Energy Analysis of Earth Sheltered Dwellings." Master's thesis, Oklahoma State University, 1980. Available from Architectural Extension, Oklahoma State University.[2]

Birkerts, Gunnar. *Subterranean Urban Systems.* Ann Arbor, Mich.: University of Michigan, 1974. Available from University of Michigan, Institute of Science and Technology, Industrial Development Division, 2200 Bonisteel Blvd., Ann Arbor, MI 48109.

Bligh, Thomas P. "Energy Conservation by Building Underground." *Underground Space* 1 (1976):19-33. Article available from Underground Space Center, University of Minnesota.[1]

Boyer, Lester L. "The Earth Sheltered Housing Phenomenon." In *Education and Industry: A Joint Endeavor,* the American Society of Engineering Education (ASEE) annual conference proceedings, Los Angeles, Calif., June 1981. Available from ASEE, 11 Dupont Circle, Suite 200, Washington, DC 20036.

_____. *Earth Sheltered Structures Fact Sheets: No. 7, Daylighting Design; No. 8, Indoor Air Quality; No. 9, Earth Coupled Cooling Techniques; No. 10, Disaster Protection; No. 11, Building in Expansive Clays; No. 12, Passive Solar Heating.* Stillwater, Okla.: Oklahoma State University, 1981. Available from Architecture Extension, Oklahoma State University.[2]

_____. "Earth Shelter Trends in the South Central Plains." In *Proceedings — Underground Space Conference and Exposition,* Kansas City, Mo., June 1981. Elmsford, N.Y.: Pergamon Press, 1981. Individual paper available from Architectural Extension, Oklahoma State University.[2]

_____. "Evaluation of Energy Savings Due to Daylighting." In *Proceedings of the AS/ISES International Passive and Hybrid Cooling Conference,* Miami Beach, Fla., November 1981. Edited by Arthur Bowen; Eugene Clark; and Kenneth Labs. Available from AS/ISES.[3]

Boyer, Lester L., ed. *Earth Sheltered Condominiums*. Stillwater, Okla.: Oklahoma State University, 1980. Available from Architectural Extension, Oklahoma State University.[2]

_____. *Proceedings — Earth Sheltered Building Design Innovations Conference,* Oklahoma City, Okla., April 1980. Available from Architectural Extension, Oklahoma State University.[2]

_____. *Proceedings — October 1981: Earth Shelter Performance and Evaluation Conference,* Oklahoma State University, 1981. Available from Architectural Extension, Oklahoma State University.[2]

Boyer, Lester L.; Grondzik, Walter T.; and Fitzgerald, Daniel K. "Earth Sheltered Housing Performance: A Summary Report." In *Proceedings of the AS/ISES Annual Meeting,* Philadelphia, Pa., May 1981. Edited by Barbara H. Glenn and Gregory E. Franta. Available from AS/ISeS.[3]

Boyer, Lester L., and Johnston, Timothy L. "Organization of Interior Spaces for Earth Cooling." In *Proceedings of the AS/ISES International Passive and Hybrid Cooling Conference,* Miami Beach, Fla., November 1981. Edited by Arthur Bowen; Eugene Clark; and Kenneth Labs. Available from AS/ISES.[3]

Boyer, Lester L.; Weber, Margaret J.; and Grondzik, Walter T. *Energy and Habitability Aspects of Earth Sheltered Housing in Oklahoma*. Stillwater, Okla.: Oklahoma State University, 1980. Available from Architectural Extension, Oklahoma State University.[2]

Brown, James J., and Gannon, David L. *Underground Homes: Primer to Earth Sheltered Living*. Portsmouth, Ohio: Underground Homes, 1979.

Brown, James J., and Johnson, Terry A. *Earth Shelter Waterproofing: The Guide to Total Moisture Control*. Portsmouth, Ohio: Underground Homes, 1980.

Calvert, Terri. "The Solartron Prefabricated Earth-Sheltered Home," *Mother Earth News,* May/June 1977, pp 157–59.

Campbell, Stu. *The Underground House Book*. Charlotte, Vt.: Garden Way, 1980.

Chalmers, Larry S., and Jones, J. *Homes in the Earth*. San Francisco, Calif.: Chronicle Books, 1980.

Clark, Kenneth N., and Paylore, Patricia, eds. *Desert Housing: Balancing Experience and Technology for Dwelling in Hot Arid Zones*. Tucson, Ariz.: University of Arizona Press, 1980.

"Colosol's Earth-Sheltered Energy Saver." *Mother Earth News,* March/April 1980, pp. 136–37.

Concrete Masonry Solar Architecture. A quarterly journal available from National Concrete Masonry Association, 2302 Horse Pen Rd., P.O. Box 781, Herndon, VA 22070.

Cook, Marcia J. "Factors Associated with the Decision to Live in an Earth Sheltered House." Master's thesis, Oklahoma State University, 1981. Available through interlibrary loan.

Crowther, Richard L.; Karius, Paul; Atkinson, Lawrence; and Frey, Donald. *Sun/Earth: How to Use Solar and Climatic Energies Today.* Denver, Colo.: Crowther/Solar Group Architects, 1976.

Dean, Andrea O. "A Summary of Underground Housings' 'How-To' Manual." *American Insitute of Architects Journal* 67 (1978):44–45.

——. "Underground Architecture." *American Institute of Architects Journal* 67 (1978):34–51.

——. "The Underground Movement Widens," *American Institute of Architects Journal* 67 (1978):34.

Dempewolff, Richard F. "Underground Housing." *Science Digest,* November 1975, p. 40.

——. "Your Next House Could Have a Grass Roof." *Popular Mechanics,* March 1977, p. 78.

Dixon, James M. "Working the Land." *Progressive Architecture,* April 1979, pp. 142–43.

"A Down-to-Earth Architect." *Mother Earth News,* January/February 1981, pp. 174–75.

"Earthform House by William Morgan Echoes Florida's Pre-Colonial Past." *Architectural Record* 159 (1976):106–7.

Earth Shelter Corporation of America. *Earth Shelter Corporation of America.* Berlin, Wis.: Earth Shelter Corp. of America, 1979. Brochure available from Earth Shelter Corp., Route 2, Box 97B, Berlin, WI 54923.

"Earth Sheltered Housing/Passive Solar Heating: Special Edition." *Alternative Sources of Energy,* August 1978, entire issue.

"Earth Sheltered Housing/Passive Solar Heating: Special Edition." *Alternative Sources of Energy,* September/October 1979, entire issue.

"An Earth Shelter for Independence." *Mother Earth News,* September/October 1980, pp. 132–34.

Earth Shelter Living magazine (previously *Earth Shelter Digest & Energy Report)* is published bimonthly by WEBCO Publishing, Inc., 1701 E. Cope, St. Paul, MN 55109.

Edelhart, Mike. "The Good Life Underground." *Omni,* January 1980, p. 50.

_____. *The Handbook of Earth Shelter Design.* Garden City, N.Y.: Doubleday and Co., 1981.

Ellison, Tom, and Carmody, John. "Earth Sheltered Houses: Two Case Studies." *Architecture Minnesota* 6 (1980):42–44.

Engelken, Ruth. "Underground in Tunisia." *Mother Earth News,* November/ December 1980, pp. 154–55.

Fairhurst, Charles. "Going Under to Stay on Top." *Underground Space* 1 (1976):71–86. Article available from Underground Space Center, University of Minnesota.[1]

Ferguson, M. *Reinforced Concrete Fundamentals.* New York: John Wiley & Sons, 1973.

Ford, Barbara. "Safe City: Apartment Living inside a Mountain." *Science Digest,* August 1969, pp. 16–19.

Forsyth, Benjamin. *Unified Design of Reinforced Concrete Members.* New York: McGraw-Hill Book Co., 1971.

"Forum: Energy's Impact on Architecture." *Architectural Forum* 139 (1973):59–74.

Geiger, Rudolf. *The Climate Near the Ground.* Cambridge, Mass.: Harvard University Press, 1965.

Goldberger, Paul. "The House in the Hill." *New York Times Magazine,* 7 August 1977, pp. 42–44.

Griffin, C. W., Jr. *Energy Conservation in Buildings: Techniques for Economical Design.* Washington, D.C.: Construction Specifications Institute, 1974.

Grondzik, Walter T. "Energy Conservation in Earth Sheltered Structures." A paper presented at the American Society of Agricultural Engineers (ASAE) annual meeting, Orlando, Fla., June 1981. Available from ASAE, 2950 Niles Rd., St. Joseph, MI 49085.

Grondzik, Walter T., and Boyer, Lester L. "Performance Evaluation of Earth Sheltered Housing in Oklahoma." In *Proceedings of the AS/ISES Annual Meeting,* Phoenix, Ariz., June 1980. Edited by Gregory E. Franta and Barbara H. Glenn. Available from AS/ISES.[3]

Grondzik, Walter T.; Boyer, Lester L.; and Johnston, Timothy L. "Earth Coupled Cooling Techniques." In *Proceedings of the AS/ISES Sixth National Passive Solar Conference,* Portland, Ore., September 1981. Edited by John Hayes and William Kolar. Available from AS/ISES.[3]

_____. "Variations in Earth Covered Roof Temperature Profiles." In *Proceedings of the AS/ISES International Passive and Hybrid Cooling Conference,* Miami Beach, Fla., November 1981. Edited by Arthur Bowen; Eugene Clark; and Kenneth Labs. Available from AS/ISES.[3]

GTE Sylvania, Inc. *Sylvania Lighting Handbook.* 3d ed. Danvers, Mass.: GTE Sylvania, Inc. 1969.

Guy, Homer L. "An Economic Comparison of Passively-Conditioned Underground Houses." Master's thesis, Oklahoma State University, 1981. Available from Architectural Extension, Oklahoma State University.[2]

Haupert, David. "Underground Housing Is Coming on Strong." *Better Homes and Gardens,* September 1979, p. 97.

Ingersoll, John. "12 Subterranean Pioneers Report: 'It's Great to Live Underground.' " *Popular Mechanics,* May 1980, p. 114.

Israel, Frank. "Architecture: William Morgan." *Architectural Digest,* December 1979, pp. 86–93.

Johnston, Timothy L. *A Tour of Selected Oklahoma Earth Shelters, II.* Stillwater, Okla.: Oklahoma State University, 1981. Available from Architectural Extension, Oklahoma State University.[2]

Kahn, Lloyd, Jr., ed. *Shelter.* New York: Random House, 1973.

Kaufman, John. *IES Lighting Handbook.* 2d ed. New York: Illuminating Engineering Society, 1954.

Keehn, Pauline A. *Earth Sheltered Housing: An Annotated Bibliography and Directory.* Chicago: Council of Planning Librarians, 1981. Available from Council of Planning Librarians, Rm. 466, 1313 E. 60th St., Chicago, IL 60637-2897. Order no. CPL Bibliography 43.

Labs, Kenneth B. *Land-Use Regulation of Underground Housing.* Chicago: American Planning Association, 1979. Available from American Planning Association, Rm. 248, 1313 E. 60th St., Chicago, IL 60637. Order no. Planning Advisory Service Memo 79.5.

_____. *Regional Analysis of Ground and Aboveground Climate.* Oak Ridge, Tenn.: Oak Ridge National Laboratory, 1981. Available from National Technical Information Service, Springfield, VA 22161. Order no. ORNL/Sub-81/40451/1.

_____. "Underground Building Climate." *Solar Age,* October 1979, p. 44.

Lambe, T. William, and Whitman, Robert V. *Soil Mechanics.* New York: John Wiley & Sons, 1979.

"Landis and Pamela Gores's Semi-Subterranean 'House for All Seasons,' " *Mother Earth News,* January/February 1978, pp. 64–65.

Lane, Charles A. "Waterproofing Earth-Sheltered Houses." *Fine Homebuilding,* April/May 1981, pp. 35–37.

Langewiesche, Wolfgang. "There's a Gold Mine under Your House." *House Beautiful,* August 1950, p. 92.

Langley, John. *Earth Sheltered Sun Belt Homes; 20 Plans and Sketches.* Winter Park, Fla.: Sun Belt Earth Sheltered Research, 1982. Available from Sun Belt Earth Sheltered Research, P.O. Drawer 729, Winter Park, FL 32790.

Langley, John, and Gay, James L. *Sun Belt Earth Sheltered Architecture.* Winter Park, Fla.: Sun Belt Earth Sheltered Research, 1980. Available from Sun Belt Earth Sheltered Research, P.O. Drawer 729, Winter Park, FL 32790.

Lorenz, Ray. "The Underground Home." *The Family Handyman,* February 1980, p. 12.

Marcovich, Sharon J. "Buried Bookstore Saves Energy, Saves Space, Saves the View." *Popular Science,* September 1977, pp. 96–97.

Martindale, David. *Earth Shelters.* New York: E. P. Dutton, 1981.

_____. "New Homes Revive the Ancient Art of Living Underground." *Smithsonian,* February 1979, p. 96.

Mason, Roy. "Beyond 2000 Architecture." *The Futurist,* October 1975, pp. 235–46.

_____. "Underground Architecture." *The Futurist,* February 1976, pp. 16–20.

Mazria, Edward. *The Passive Solar Energy Book.* Emmaus, Pa.: Rodale Press, 1979.

Moreland, Frank L., ed. *Alternatives in Energy Conservation: The Use of Earth Covered Buildings.* Washington, D.C.: U.S. Government Printing Office, 1975. Order no. 038-000-00286-4.

Moreland, Frank L.; Higgs, Forrest; and Shih, Jason, eds. *Earth Covered Buildings: Technical Notes.* Springfield, Va.: National Technical Information Service, 1978. Order no. Conf. 7805138-P1.

Morgan, William. "Molding Our Man-Made World." *American Institute of Architects Journal* 61 (1974):32–39.

———. "Up to Earth." *Progressive Architecture,* April 1979, pp. 84–87.

Morgan, William N. "Buildings as Landscape: Five Current Projects by William Morgan." *Architectural Record* 152 (1972):129–36.

Morton, David. "Esprit Grows in Brooklyn." *Progressive Architecture,* May 1978, pp. 62–67.

———. "The Solar Underground." *Progressive Architecture,* April 1979, pp. 124–27.

Murphy, Jim. "Cooperate with Nature." *Mother Earth News,* November/December 1980, pp. 114–16.

———. "House in the Hill." *Progressive Architecture,* October 1980, pp. 72–75.

Neal, Wallace. "Human Spaces Underground — An Old/New Technology." *The Construction Specifier* 31 (1978):39–50.

Newman, Jerry; Godbey, Luther C.; and Davis, Martin A. *Design Considerations for Below-Grade-Level Houses.* A paper presented at the American Society for Agricultural Engineering (ASAE) Winter Meeting, Chicago, Ill., December 1978. Available from ASAE, 2950 Niles Rd., St. Joseph, MI 49085.

Oehler, Mike. *The Fifty Dollar and Up Underground House Book.* Bonners Ferry, Idaho: Mole Publishing Co., 1978.

Olgyay, Victor. *Design with Climate.* Princeton, N.J.: Princeton University Press, 1963.

Parizek, E. J. *Development and Utilization of Underground Space.* Springfield, Va.: National Technical Information Service, 1975. Order no. PB252812.

Portland Cement Association. *Constructing Earth Sheltered Housing with Concrete.* Skokie, Ill.: Portland Cement Association, 1981. Available from Portland Cement Association, 5420 Old Orchard Rd., Skokie, IL 60077.

———. *Earth-Integrated Building Construction.* Skokie, Ill.: Portland Cement Association, 1978. Available from Portland Cement Association.

———. *Earth-Sheltered Concrete Homes.* Skokie, Ill.: Portland Cement Association, 1980. Available from Portland Cement Association.

"Private Residence." *Architectural Record* 167 (1980):114–16.

Ramsey, C. G., and Sleeper, H. R. *Architectural Graphic Standards.* 6th ed. New York: John Wiley & Sons, 1970.

Rinker, D. "Go Underground in Michigan." *Mother Earth News,* November 1978, pp. 112–13.

Rivers, W. Joel; Helm, Bob; Warde, William D.; and Grondzik, Walter T. "Analysis of Earth Sheltered Dwellings in the South Central States." In *Proceedings of the AS/ISES International Passive and Hybrid Cooling Conference,* Miami Beach, Fla., November 1981. Edited by Arthur Bowen; Eugene Clark; and Kenneth Labs. Available from AS/ISES.[3]

Rivers, W. Joel; Warde, William D.; and Helm, Bob. "A Comparison of Assessments by Above Ground and Earth Shelter Occupants." In *Proceedings– October 1981: Earth Shelter Performance and Evaluation Conference,* Oklahoma State University, 1981. Edited by Lester L. Boyer. Proceedings and individual paper available from Architectural Extension, Oklahoma State University.[2]

Rollwagen, Mary. "Is Earth Sheltered Housing Entering the Mainstream?" *Architecture Minnesota* 6 (1980):29–30.

Roy, Robert L. "A Pair of Cordwood-Construction Pioneers Have Gone Underground, in . . . a Log-End Cave." *Mother Earth News,* January/February 1981, pp. 110–111.

_____. *Underground Houses: How to Build a Low-Cost Home.* New York: Sterling Publishing Co., 1979.

Rush, Richard. "Innovations in Concrete." *Progressive Architecture,* May 1978, pp. 100–109.

"Saving by Going Underground." *American Institute of Architects Journal* 61 (1974):48–49.

Scalise, James, ed. *Earth Integrated Architecture.* Tempe, Ariz.: Arizona State University, College of Architecture, 1975.

Scott, Ray. *How to Build Your Own Underground Home.* Blue Ridge Summit, Pa.: TAB Books, 1979.

_____. *Underground Homes: An Alternative Lifestyle.* Blue Ridge Summit, Pa.: TAB Books, 1981.

Seeley, Barrett, ed. *Earth-Sheltered Housing: The Comprehensive Bibliography.* Washington, D.C.: Eco-Terra, 1981. Available from Eco-Terra, 1710 Connecticut Ave. NW, Washington, DC 20009.

Smay, V. Elaine. "Popular Science Leisure-Home Plan: Underground Solar House." *Popular Science,* December 1978, pp. 86–87.

_____. "Solar Goes Underground in New House Designs." *Popular Science,* May 1980, pp. 82–88.

_____. "Underground Houses — Low Fuel Bills, Low Maintenance, Privacy, Security." *Popular Science,* April 1977, p. 84.

_____. "Underground Living in this Ecology House Saves Energy, Cuts Building Costs, Preserves the Environment." *Popular Science,* June 1974, p. 88.

Smith, Ronald. *Principles and Practices of Light Construction.* 2d ed. Englewood Cliffs, N.J.: Prentice-Hall, 1970.

Stauffer, Truman, Sr. *Occupancy and Use of Underground Mined Out Space in Urban Areas: an Annotated Bibliography.* Chicago: Council of Planning Librarians, 1974. Available from Council of Planning Librarians, Rm. 466, 1313 E. 60th St., Chicago, IL 60637-2897. Order no. CPL Exchange Bibliography 602.

Sterling, Raymond, and Larson, Nancy E., eds. *Going Under to Stay on Top — Housing.* 1978–1979 Earth-Sheltered Conference Series transcripts of selected presentations. Available from Underground Space Center, University of Minnesota.[1]

Tingerthal, Mary. "Deciding to Build an Earth Sheltered Home." *Architecture Minnesota* 6 (1980):36–38.

_____. "MHFA Solar/Earth Sheltered Demonstration Housing Program: Seven Case Studies." *Architecture Minnesota* (1980):45–48.

Underground Heat and Chilled Water Distribution Systems. Washington, D.C.: U.S. Government Printing Office, 1975. Order no. 003-003-01417-6.

Underground Space Center, University of Minnesota. *Earth Sheltered Community Design: Energy-Efficient Residential Development.* New York: Van Nostrand Reinhold, 1981.

_____. *Earth Sheltered Homes: Plans and Designs.* New York: Van Nostrand Reinhold, 1981.

_____. *Earth Sheltered Housing: Code, Zoning, and Financing Issues.* Washington, D.C.: U.S. Government Printing Office, 1980. Order no. 023-000-00632-4.

_____. *Earth Sheltered Housing Design: Guidelines, Examples, and References.* New York: Van Nostrand Reinhold, 1979.

"Underground — the Prairie Sod House Returns." *Architecture Minnesota* 3 (1977):24–31.

von Fraunhoffer, Herman J. *The Housing Alternative.* Glendale, Ariz.: Concept 2000, 1979. Available from Concept 2000, Inc., 19003 N. 52nd Ave., Glendale, AZ 85308.

Wade, Alex, and Ewenstein, Neal. *30 Energy-Efficient Houses . . . You Can Build.* Emmaus, Pa.: Rodale Press, 1977.

Wampler, Jan. *All Their Own: People and Places They Build.* Cambridge, Mass.: Schenkman Publishing Co., 1977.

Wampler, Louis. *Underground Homes.* Rev. ed. New York: Pelican, 1980.

Wells, Malcolm. *Energy Essays.* Barrington, N.J.: Edmund Scientific Co., 1976.

_____. "Environmental Impact." *Progressive Architecture,* June 1974, pp. 59–63.

_____. *Gentle Architecture.* New York: McGraw-Hill Book Co., 1981.

_____. "Underground Architecture." *CoEvolution Quarterly,* Fall 1976, pp. 84–93.

_____. *Underground Designs.* Brewster, Mass.: Malcolm Wells, 1977. Available from Malcolm Wells, Box 1149, Brewster, MA 02631.

Wells, Malcolm, ed. *Notes from the Energy Underground.* New York: Van Nostrand Reinhold, 1980.

"Winston House, Lyme." *Architectural Record* 155 (1974):52–53.

Wynne, George B. *Reinforced Concrete Structures.* Reston, Va.: Reston Publishing Co., 1981.

Young, H. W.; Wright, R. R.; Swenson, R. W.; Stone, A. W.; and Hoch, I. *Legal, Economic and Energy Considerations in the Use of Underground Space.* Springfield, Va.: National Technical Information Service, 1974. Order no. PB236755.

1. Underground Space Center
   University of Minnesota
   11 Mines and Metallurgy Bldg.
   221 Church St. SE
   Minneapolis, MN 55455
2. Architectural Extension
   Oklahoma State University
   120 Architecture Bldg.
   Stillwater, OK 74078
3. American Section/International Solar Energy Society (AS/ISES)
   110 W. 34th St.
   New York, NY 10001

Photograph Credits

Photo 1-1: Richard Kennedy, designer/builder, Burnsville, N.C.

Photo 1-2: Robert S. Cole, The Evergreen State College, Olympia, Wash.

Photo 1-3: Heinz Stefan, University of Minnesota

Photo 1-4: Greg Stanford, architect

Photo 1-5: Great Midwest Corporation, Kansas City, Mo.

Photo 1-6: Michael Reese Much, Phoenix, Ariz.

Photo 1-7: Lon B. Simmons, Simmons & Sun, Inc., High Ridge, Mo.

Photo 2-1: Mitchell T. Mandel, Rodale Press

Photo 2-2: Herb Wade

Photo 2-3: Herb Wade

Photo 2-4: Architectural design and photography by Richard M. Sibly, A.I.A., Atlanta, Ga.

Photo 2-5: Robert S. Cole, The Evergreen State College, Olympia, Wash.

Photo 2-6: Hawkins Photography, Hyannis, Mass.

Photo 2-7: Mitchell T. Mandel, Rodale Press

Photo 2-8: Caterpillar Tractor Co., Peoria, Ill.

Photo 2-9: Mitchell T. Mandel, Rodale Press

Photo 2-10: Mitchell T. Mandel, Rodale Press

Photo 2-11: Herb Wade

Photo 3-1: Carl Doney, Rodale Press

Photo 3-2: Carl Doney, Rodale Press

Photo 3-3: T. L. Gettings, Rodale Press

Photo 3-4: Carl Doney, Rodale Press

Photo 3-5: Mitchell T. Mandel, Rodale Press

Photo 3-6: Carl Doney, Rodale Press

Photo 3-7: T. L. Gettings, Rodale Press

Photo 3-8: Edmund Scientific Company, Barrington, N.J.

Photo 3-9: Dalen Products, Inc., Knoxville, Tenn.

Photo 3-10: David Yamamoto

Photo 3-11: Carl Doney, Rodale Press

Photo 3-12: Mitchell T. Mandel, Rodale Press

Photo 4-1: Reprinted with the permission of *Earth Shelter Living* magazine

Photo 4-2: T. L. Gettings, Rodale Press

Photo 4-3: Ingersoll-Rand Compaction Division, Shippensburg, Pa.

Photo 4-4: USDA Soil Conservation Service

Photo 5-1: T. L. Gettings, Rodale Press; designed by Shelter Design Group and built by Consolarnation, Stony Run, Pa.

Photo 5-2: Portland Cement Association, Skokie, Ill.

Photo 5-3: Lon B. Simmons, Simmons & Sun, Inc., High Ridge, Mo.

Photo 5-4: Lon B. Simmons

Photo 5-5: Lon B. Simmons

Photo 5-6: Lon B. Simmons

Photo 5-7: Bob Scott, Scott Earth Homes, Coeur d'Alene, Idaho

Photo 5-8: Architerra, Inc. Arlington, Va.

Photo 5-9: Post-Tensioning Institute, Phoenix, Ariz.

Photo 5-10: Lon B. Simmons

Photo 5-11: Roger O'Brien

Photo 5-12: Architectural design and photography by Integrated Building Systems, Grand Haven, Mich.

Photo 5-13: Architectural design and photography by Integrated Building Systems, Grand Haven, Mich.

Photo 5-14: Mitchell T. Mandel, Rodale Press

Photo 5-15: James Zanetto, architect, Davis, Calif.

Photo 6-1: W. R. Grace & Co., Cambridge, Mass.

Photo 6-2: T. L. Gettings, Rodale Press

Photo 6-3: Effective Building Products Inc., Cleveland, Ohio

Photo 6-4: American Colloid Company, Skokie, Ill.

Photo 7-1: Reprinted with the permission of *Earth Shelter Living* magazine

Photo 7-2: Carl Doney, Rodale Press

Photo 7-3: David Sellers, research engineer, Product Testing Department, Rodale Press

Photo 7-4: Ted First, designer/builder, Pennsylvania Furnace, Pa.

Photo 8-1: Margaret Smyser, Rodale Press

Photo 8-2: John Cooper

Photo 8-3, 8-4: Margaret Smyser, Rodale Press

Photos 8-5, 8-6, 8-7: Carl Doney, Rodale Press

Photos 8-8, 8-9, 8-10, 8-11, 8-12, 8-13, 8-14, 8-15, 8-16, 8-17, 8-18, 8-19, 8-20: Mitchell T. Mandel, Rodale Press

Photos 8-21, 8-22, 8-23, 8-24: Michael Reese Much, Phoenix, Ariz.

Photos 8-25, 8-26, 8-27: Mitchell T. Mandel, Rodale Press

Photos 8-28, 8-29, 8-30, 8-31, 8-32: Stephen O. Muskie, Biddeford, Maine

Index

A

absorption, water, and soil particles, 82–84
adhesives, building
    for exterior insulation, 168–69
    panel cement as, 144
admixtures, for concrete, 103–4
aggregate, size and grading of, 105
air cleaners, filtering out dust with, 202–3
air conditioners
    controlling humidity levels with, 194–95
    disadvantages of, 195
air deodorizers
    filtering out odors with, 203–4
    houseplants as, 204
    kitchen stove hood as, 203
air-entraining agents, added to concrete, 104
air-exchange rate, uncontrolled infiltration and, 200–201
air pollutants
    sources of, 200
    ventilation for, 199–200
air purifiers, for underground houses, 202–4
air registers, placement of, 67–68, 70
All-Weather Wood Foundation System. *See* wood construction
American Society of Professional Engineers (ASPE), as source for structural engineers, 99
aquifer, 84–85
    for defining saturation levels, 85
armature, wire-mesh, used in ferrocement construction, 132–34
artificial lighting
    color of, 212
    underground houses and, 210–12
asphalt emulsion, as waterproofing material, 152
atrium, 22–25, 71–72, 210
    design of, 23
    dome as cover for, 24
    drains for, 23
    improving solar utilization with, 71–72
    isolated-gain system and, 71–72
    as lighting source, 210
    wind conditions and, 24
attached solar greenhouse. *See also* solar greenhouse
    controlling environment of, 65
    as heat source, 65–66
    isolated-gain system and, 59, 64–66
    roof, 66
auxiliary heating systems, for underground houses, 185–89

B

baseboard heaters, as auxiliary heating system, 186–87
beadboard, as exterior insulation, 171
bendable plastic pipe, used in plumbing, 180–81
bentonite
    precautions for use of, 163
    as waterproofing material, 84, 160–61
Bentonize, as waterproofing material, 161
Bernold sheets, application of concrete to, 134
block construction, 137–43. *See also* concrete block construction
block structures, waterproofing of, 161–62
block walls, increasing strength of, 142
bond-beam, as form of block construction, 139
building adhesives
    for exterior insulation, 168–69
    panel cement as, 144
building sites
    designing for, 97–98
    flat, 18–22
    subsurface analysis of, 98–99

Page numbers in boldface indicate table or chart information.

277

building sites *(continued)*
> for underground houses, 16–18
> water removal from, 19

C

calcium chloride, as concrete hardening agent, 103
capillary action, soil and, 82–84
cast-in-place concrete, 118–19
> cold joint and, 118
> pouring types of, 118–19
> waterstops for, 119
cast-iron pipes, used in plumbing, 183
caves, living in, 3
CCA-treated wood, for roofs, 149
cementitious products, for water-proofing, **158–59**
> manufacturers of, 259
Characteristics of Concrete Admixtures, **104**
Characteristics of Plastic Piping, **181**
chromated copper arsenate (CCA), as wood preservative, 145, 147–48
Clear-Day Solar Heat Gain for Three Latitudes, **40**
clerestory windows. *See also* windows
> advantages of, 16
> as light source, 209
climate
> matching cooling system to, 196
> surface, 93
> > around a building, 94–96
> underground, 87, 93–94
> > soil's effect on, 88
clouds, effect on sunlight, 35–36
cold joint, cast-in-place concrete and, 118
color, interior, and storage mass, 64
comfort control, earth sheltering and, 4, 192–99
compaction, soil and, 79, 81–82
> moisture needed for, 82
compression strength, soil and, 79, 81

concealed wiring, for underground houses, 178–79
concrete
> admixtures for, 103–4
> > characteristics of, **104**
> advantages of, 100
> aggregate, 103, 105
> air-entraining agents for, 104
> application to wire mesh, 134–36
> best days for pouring, 109
> block, 137–43
> calcium chloride and, 103
> cast-in-place, 118–19
> characteristics of, 104–5
> components of, 103
> curing process for, 109
> determinant for strength of, **105**
> determining thickness of, 117
> ferrocement construction, 131–36
> forms for, 119–20
> hardening agents for, 103
> hydration of, 104–5
> mixing process of, 105–7
> monolithic pour, 118
> pouring of, 108–9
> precast panels, 123–26
> prestressed, 120–29
> rebar for, 108, 112
> reinforcing, 111–15
> > grades and strengths, **112**
> > proper placement of, 113
> > tension, 114
> sectional pour, 119
> shear stress on, 115–16
> sprayed, 134–35
> structural characteristics of, 110–11
> temperature reinforced, 114–15
> tension reinforced, 111–15
> > problems in designing, 112–15
> timing the pour, 108
> varied strength of, **106**
> waterproofing of, 100
concrete admixtures, 103–4

concrete block construction, 137–43
    bond-beam, 139
    composition of, 137–38
    increasing wall strength with,
        142
    rebar for, 139
    reinforcement of, 138–41
        grout for, 138–39
    roofs, 142–43
    surface-bonded, 141–42
    types of, 139, 141–42
    waterproofing of, 161–62
concrete floor, as thermal storage,
    62
condensation, on windows, 52
conduit, for underground electrical
    systems, 175–78
    bending tools for, 176
    installation of, 175–78
    National Electrical Code and,
        176
constant-temperature depth
    determining of, 93–94
    effect on amount of insulation,
        165
construction, concrete, types of,
    110–43. *See also* construction
    techniques
construction techniques
    concrete, 100–143
        block, 137–43
        cast-in-place, 118–20
        ferrocement, 131–36
        posttensioned, prestressed,
            126–30
        prestressed, 120–23
        pretensioned, prestressed,
            123–26
    earth contact, 143–47
    wood, 147–51
cooling systems, for underground
    houses, 192–99
cool-tube systems
    cooling underground houses
        with, 195–98
    determining rates of air flow
        through, 196
    drainage and, 195–96

fans and, 197
installation precautions for, 198
Cooper house, 214–19
    concrete block construction,
        214
    construction details for, 216,
        219
    cooling, 218
    floor plan of, 214
    heating, 215, 218
    insulation, 216
    plumbing, 218
    roof construction, 218
    ventilation, 215–16, 218
    walls, 214–15, 218
    waterproofing, 216
    wiring, 218
curing process, of concrete, 109

D

daylighting, underground houses
    and, 208–10. *See also* sunlight
dehumidifiers, controlling humidity
    levels with, 194
diffusing skylight, as light source,
    208–9
direct-gain systems, 42–53, 75
    glare and, 49–50
    guidelines for, 43–50
    improving performance with,
        50–53
    reducing heat loss and, 50–53
    room color and, 49
    shading methods for
        roof overhang, 53
        trees, 53
    storage mass for, 42–43
    summer operation of, 53
    temperatures with, 48–49
    windows for, 42–45
domestic hot water systems, for un-
    derground houses, 72–74
drain(s)
    for atriums, 23
    French, 13
    gutters as, 15–16

drain(s) *(continued)*
  sewage disposal and grading of, 183
  subsurface, 154–55
drainage, 86, 152–56
  cool tubes and, 195–96
  Enkadrain for, 155–56
  French drain for, 13
  gutters for, 15–16
  roof as continuation of hillside slope for, 16
  sloping sites and, 12–18
  subsurface drains for, 154–55
  swale method for, 13, 15
  tiles for, 19–20, 98–99, 154–55
drainage tiles, 19–20, 154–55
  placement of, 98–99
dust
  effect on sunlight, 36
  filtering out, 202–3

E

earth-contact houses, 20–21, 143–47
  advantages of, 143, 145–46
  disadvantages of, 143, 146
  energy efficiency of, 143
  foam insulation for, 143
  insulation for, 171
  site for, 20–21
  waterproofing of, 145
earth's axis, effect on solar design, 32–33
earth sheltering. *See also* underground living
  advantages of, 4–5
  comfort control and, 4, 192–99
  disadvantages of, 6–8
  economic security and, 4–5
  erosion problems and, 18
  insulation and, 5
  internal heat generation and, 5
  physical security and, 4
  reasons for, 3
  sound control and, 5
  water control and, 12–18

economic security, earth sheltering and, 4–5
economizer, as ventilation subsystem, 201–2, 204
electrical systems
  concealing wiring of, 178–79
  conduit for, 175–78
  installation problems with, 175–77
  National Electrical Code and, 176
  surface-mounted wiring devices, 177
  for underground houses, 174–79
emergency exits, floor plan design and, 28–29, 150
energy efficiency, earth-contact houses and, 143
Enkadrain, for water drainage, 155–56
erosion problems, for earth-sheltered houses, 18
exterior insulation, 168–69, 171, 173
  beadboard for, 171
  disadvantages of, 168
  installation of, 169
  keeping in place, 168–69
  minimum thickness for, 173
  preferred material for, 173

F

fan
  determining flow rates with, 197
  for isolated-gain system, 68–71
  multi-speed, 70–71
  variable-speed
    automatic control of, 69
    changing air with, 68–70
    proportional controller for, 69
ferrocement construction, 131–36, 171–72
  advantages with, 132–33, 136
  Bernold sheets for, 134

insulation for, 171–72
uses for, 136
using rebar with, 132–33,
    143–44
wire-mesh armature for, 132,
    133–34
fiberglass, as interior insulation, 169
fire exits, planning for, 28–29, 150
flat sites, houses for, 18–22
    disadvantages of, 20–22
    water removal from, 19
floor plans, layout of, 27–31
    emergency exits and, 28–29,
        150
    plumbing system and, 30
flow rates, fans to determine, 197
Foamular (pinkboard), as insulation
    material, 171
Freidrichs house, 220–24
    construction details for, 223
    cooling, 223
    floor plan of, 220
    heating, 223
    insulation, 223
    plumbing, 223
    roof construction, 223
    ventilation, 223
    walls, 221, 223
    waterproofing, 223
    wiring, 223
    wood construction, 221
French drain, for water drainage, 13

G

geothermal energy, effect of soil
    temperature on, 88
glazing
    cleaning of, 57
    for indirect-gain systems, 53,
        55–57
    materials for, 57
    requirements for, 60
    of sunspace, 60, 63
    thermal storage for, 60–63
grading, of soil, 85

gravity water, waterproofing prob-
    lems and, 84, 86
greenhouse. See solar greenhouse
grout
    bonding steel to concrete with,
        126
    reinforcement for concrete
        blocks, 138–41
gutters
    for drainage, 15–16
    removing water from, 19

H

heat
    distribution of, 189–91
    estimating requirements for,
        255–58
    internal heat generation and, 5
    solar greenhouse as source for,
        65–66
    systems for, 185–89, 191–92
    wood, 187–91
heaters, baseboard, as auxiliary heat-
    ing system, 186–87
heating degree-days (HDD), explana-
    tion of, 45
heating systems, 185–92
    auxiliary, 185–89
        baseboard heaters, 186–87
        wood stoves as, 187–89
    internal gains and, 187
heat loss, estimation of, 255
heat pump, for cooling, 195
Home Heating Index (HHI), 52, 256
    determination of, **257**
    movable insulation as estimating
        factor for, 52
hot water systems
    domestic, 72–74
    passive, 73–74
    Zomeworks' Big Fin collectors
        for, 74
houseplants, as room deodorizer,
    204

houses, design of, and underground
     site for, 9–10
          flat site for, 18–22
          sloping site for, 10–18
houses, underground. *See* under-
     ground houses
humidity, levels of, 192–95
          air conditioners for, 194–95
          dehumidifiers for, 194
          reduction of, 192, 193–95
hydration
          of concrete, 104–5
          water needed for, 105

I

illumination, efficiency of, **211.** *See
     also* lighting
indirect-gain systems, 53–59, 75
          dark surface color for, 56
          designing for, 54–56
          glazing and, 53, 55–57
          implementation of, 56
          improving performance of, 59
          movable insulation for, 59
          problems with, 53
          selective surfaces for, 56–57
          Trombe wall and, 55–56
infiltration
          air-exchange rate and, 199–201
          as ventilation, 199–201
insulation
          adding mass with, 170–71
          amount of, 172–73
          constant-temperature depth
               and, 165
          exterior, 168–69, 171, 173
               beadboard for, 171
               disadvantages of, 168
               installation of, 169
               keeping in place, 168–69
               polystyrene for, 171
          ferrocement construction and,
               171–72

fiberglass as, 169
indirect-gain system and, 59
interior, 169–71, 173
          adding mass with, 170–71
maximum levels of, 5
movable, to reduce heat loss,
     50–53
Nightwall as, 53
panels, holding in place, 144
polystyrene for, 171
R-value of, 172–73
soundproofing walls with,
     29–30
types of, 171
underground houses and,
     164–73
urethane foam as, 172
insurance, underground houses and,
     101–2
interior block walls, increasing
     strength of, 142
interior insulation, 169–71, 173
          adding mass with, 170
          R-value for, 173
internal gains, through cooking, 187
isolated-gain systems, 59–71, 75–76
          air registers for, 67–68, 70
          atrium for, 71–72
          fans for, 68–71
               automatic control of, 69
               proportional sensor for,
                    69–70
               temperature sensor for, 69
          glazing requirements for, 60
          guidelines for, 60
          skylights for, 68, 71
          solar greenhouses as, 59, 64–66
               controlling environment of,
                    65
               as heat source, 65–66
               interior color of, 64
               roof, 66
          sunspace, 60–63
               glazing requirements for,
                    60, 63
               thermal storage for, 60–63

J

Juenemann house, 225–29
   construction details for, 226–27
   cooling, 227, 229
   earth-contact, 225
   floor plan of, 225
   heating, 226–27, 229
   insulation, 226–27
   plumbing, 227
   roof construction, 227
   ventilation, 227
   walls, 225–27
   waterproofing, 226
   wiring, 227

K

kitchen stove hood, as air deodorizer, 203

L

Laughlin house, 230–34
   construction details for, 232–33
   cooling, 233
   floor plan of, 230
   heating, 233–34
   insulation, 232
   plumbing, 233
   roof construction, 230, 232
   Terra-Dome construction, 230
   ventilation, 233
   walls, 230–32
   waterproofing, 232
   wiring, 232
lighting, 206–12
   artificial, 210–12
   atrium, as source for, 210
   color contrast for, 206, 212
   daylighting, 208–10
   design error and, 207–8
   light shafts for, 209
      sources of, 210
      Wade Theory of Comfortable
         Illumination and, 206
      window placement for, 207–8
liquid-applied membranes, for waterproofing, 156–57, **158–59**
   disadvantages of, 157
   manufacturers of, 259–61

M

Matching Cooling Systems to Climates, **196**
Miley house, 235–40
   construction details for, 237,
      239
   cooling, 235, 239
   floor plan of, 235
   heating, 235, 239
   insulation, 237
   lighting, 239
   plumbing, 239
   roof construction, 237
   two-story underground house,
      235
   ventilation, 235, 239
   walls, 235–36, 239
   waterproofing, 237
   wiring, 239
modified bitumens, as waterproofing
   materials, 157
monitor skylight, reducing heat loss
   with, 71
montmorillonite clays, for waterproofing, **158–59,** 160–61
   manufacturers of, 261
mortar, for reinforcing concrete, 105
movable insulation
   as estimating factor for Home
      Heating Index, 52
   indirect-gain system and, 59
   reducing heat loss with, 50–53
   securing panels of, 52–53

N

National Electrical Code, on wiring, 176
Nightwall, as insulation mounting system, 53

P

passive hot water systems, 73–74
passive solar system designs
    advantages of, 37
    atrium as, 71–72
    direct-gain, 42–53, 75
    housing site for, 38–39, 41
    indirect-gain, 53–59, 75
    isolated-gain, 59–71, 75–76
    skylights as, 71–72
    storage mass for, 37–38, 42–43
    types of, 41–74
Pearcey house, 241–45
    building codes and, 243
    construction details for, 244
    cooling, 242, 244
    financing for, 243
    floor plan of, 241
    heating, 244
    hogan construction, 241
    insulation, 242, 244
    plumbing, 244
    roof construction, 244
    ventilation, 244
    walls, 244
    waterproofing, 242, 244
    wiring, 244
permeability, of soil, 84
physical security, earth sheltering and, 4
pipes
    cast-iron, 183
    insulation of, 180
    plastic, 180–84
        bendable, 180–81
        characteristics of, **181**
        rigid, 182
plumbing, 179–85
    cast-iron pipes for, 183
    insulating pipes, 180

    plastic pipe for, 180–84
        bendable, 180–81
        characteristics of, **181**
        rigid, 182
    pressure reducer for, 180–81, 185
    sewage disposal with, 182–84
    vents and, 184
    waterless toilets and, 184–85
    water pressure and, 180–81
pollutants, air
    sources of, 200
    ventilation for, 199–200
polybutylene pipe, for plumbing, 180
polystyrene, as exterior insulation, 171
portland cement. *See* concrete
posttensioned construction, for underground houses, 130–31
precast concrete panels, 123–26, 128–30
    construction with, 129–30
    cost-saving with, 123
    disadvantages of, 125–26, 129
    owner placement of, 128–29
preformed sheet membranes, for waterproofing, 157, **158–59,** 160
    manufacturers of, 261
    temperature for installation of, 160
preservative, for wood, 145, 147–48
prestressed concrete, 120–29
    common approaches for, 121, 123
    posttensioned, 126–28
        installing tendons for, 126
    pretensioned, 123–26
primary storage mass, for direct-gain systems, 43
Properties of Deformed Reinforcing Bars, **111**

R

radon, hazard of super-tight houses, 204
rebar. *See* steel reinforcing rods

registers, air, and placement of, 67–68

residual soil, underground construction and, 77

rigid plastic pipe, used in plumbing, 182

Rogers house, 246–49
building codes and, 248
construction costs, 246
construction details for, 247–48
financing for, 248
floor plan of, 246
heating, 247–48
insulation, 247
plumbing, 247
roof construction, 247
ventilation, 248
walls, 246–47
waterproofing, 247
wiring, 247

roll goods, for waterproofing, **160–61**
manufacturers of, 262

roof construction
block, 142–43
CCA-treated wood for, 149
pouring concrete for, 140

roof overhang, as shading method, 53

room deodorizer, houseplants as, 204

S

saturation temperature, determining of, **193**

secondary storage mass, for direct-gain systems, 43

security, underground living and, 4–5

selective surfaces, for indirect-gain systems, 56–57

sewage disposal
grading drains for, 183
plastic pipe for, 182
sewer line, slope of, 182
vents for, 184

sewer line, minimum fall needed, 182

shading
roof overhang for, 53
trees for, 53

shear strength, movement of soil and, 79, 99

shear stress, 115–17
reduction of, 116
reinforcing required for, 117
structural problems and, 115–16

Shepard house, 250–54
construction details for, 253
cooling, 253
floor plan of, 250
heating, 250–51, 253
insulation, 253
plumbing, 253
posttensioned concrete construction, 250
roof construction, 253
ventilation, 253
walls, 250–51, 253
waterproofing, 253
wiring, 253

silicone rubber, keeping insulation in place with, 168–69

sites, building
designing for, 97–98
flat, 18–22
disadvantages of, 21–22
drainage tiles for, 19–20
subsurface analysis of, 98–99
for underground houses, 16–18
water removal from, 19

skylight(s)
diffusing, 208–9
disadvantages of, 71
for isolated-gain systems, 71
monitor, 71
use of, 68

sloping sites, houses for, 10–18
advantages of, 10, 12
drainage problems and, 12–13
controls for, 13, 15
surface-water control and, 12–13, 15–16

soil
  capillary action and, 82–84
  characteristics of, 76
  chemical classification of, 76–78
  compaction of, 79, 81–82
  components, **80–81**
  composition of, 86
  compression strength and, 79,
    81
  density of, 86–87
  grading of, 85
  organic content of, 86
  particle sizes of, 78–79
  permeability of, 84
  residual, 77
  saturation of, 84–85
  shear strength, 79
  site drainage and, 86
  temperature of, 88, 92, 165,
    167–68
  thermal characteristics of, 89–93
  thermal conductivity of, 89
  types of, 78–79
  water absorption and, 82–84
  water content, 78, 82, 91
  water table and, 84–85
Soil Components and Fractions,
  80–81
soil horizons, grading and, 85
soil temperature
  effect of vegetation on, 88, 92
  effect on underground climate,
    165, 167–68
  geothermal energy and effects
    on, 88
solar chimneys, for cooling, 198–99
solar collector, room surface as, 49
solar designs
  active, 37
  effect of earth's tilt on, 32–33
  passive, 37–74
solar energy
  atriums for, 71–72
  effect of earth's tilt on, 32–33
  reducers of, 35–36
  sources of, 32–37

solar greenhouse
  controlling the environment of,
    65
  as heat source, 65–66
  interior color of, 64
  as isolated-gain system, 59,
    64–66
  on roof, 66
solar heat
  gain for, **40**
  storage of, 53
solar water-heating systems, for underground houses, 72–74
sound control, earth sheltering and,
  5
spectral distribution, of sunlight, **33**
sprayed concrete, application of,
  134–35
steel reinforcing rods (rebar)
  ferrocement construction and,
    132–33
  proper placement of, 113,
    117–18
  use for concrete block con-
    struction, 139
  use in concrete, 108, 111–15
storage mass
  direct-gain system and, 42–43
  passive system and, 37–38
  primary, 43
  secondary, 43
structural constraints, underground
  living and, 9
stucco, as building cover material,
  144–45
Styrofoam (blueboard), as insulation
  material, 171
Styrofoam Brand Insulation Mastic
  #11
  as insulation adhesive, 144,
    168–69
subsurface
  drains for, 154–55
  site analysis of, 98–99
summertime comfort control,
  192–99
  air conditioning for, 194–95
  cool tubes for, 195–98

dehumidifiers for, 194
heat pump for, 195
reducing humidity levels, 192, 194
solar chimney, 198–99
sunlight, 32–37
effect of clouds on, 35–36
effect of dust on, 36
effect of gases on, 35
spectral distribution of, **33**
surface-bonded, as form of block construction, 141–42
surface-mounted wiring, devices for, 177–78
surface-water control
approaches for, 12–13, 15–16
drainage methods for, 13, 15–16, 98–99, 154–55
swale method, for drainage, 13, 15

T

temperature, variation with depth, **90**
temperature reinforcement, used in concrete, 114–15
temperature sensor, for variable-speed fans, 69
tension reinforcement
proper placement of, 113
use in concrete, 108, 111–15
thermal conductivity, of soil, 89
thermal flywheel, effect on temperature, 164–65, 170
thermal storage, concrete floor as, 62
thermosiphon
creation of, 56, 67
glazing for, 56
manual control of, 70–71
passive hot water systems and, 73–74
tiles, drainage
placement of, 98–99
for water control, 19–20, 154–55
toilets, waterless, in underground houses, 184–85

trees, as shading method, 53
Trombe wall, indirect-gain system and, 55

U

underground environment
designing for, 97–99
subsurface site analysis, 98–99
underground houses
auxiliary heating systems for, 185–89
cooling systems for, 192–99
electrical systems for, 174–78
emergency exits for, 28–29, 150
estimating heating needs for, 256, 258
heating systems for, 185–89, 191–92
insurance for, 101–2
lighting for, 206–12
plumbing for, 179–85
sites for, 16–18
solar solutions for, 32–37
underground living. *See also* earth sheltering
advantages of, 4–5
climate and, 87, 93–94
disadvantages of, 6–8
lighting for, 206–12
physical security and, 4
plumbing for, 179–85
reasons for, 3
structural constraints and, 9
urethane foam, as insulation material, 172

V

vegetation, effect on soil temperature and, 88, 92
ventilation, 196–205
air cleaners for, 202
air-exchange rate and, 200–201
controlling humidity levels with, 194

ventilation *(continued)*
    designing systems for, 204–5
    determining needs for, 200–201
    economizers, 201–2, 204
    goal of, 199–200
    indoor pollutants and, 199–200
    infiltration and, 199–201
    reasons for, 196
ventilation systems
    designing of, 204–5
    economizer for, 201–2, 204
vents, for sewage disposal, 184
Volclay, as waterproofing material,
    161

W

walls, soundproofing of, 29–30
water
    capillary action of, 82–84
    content in soil, 86, 91
    drainage of, 86
    gravity and, 84, 86
    removal from site, 19
    table, 84–85, 98
water control, for earth-sheltered
    houses, 12–18
    drainage tiles for, 19–20, 98–99,
        154–55
water pressure
    checking of, 180–81
    reducers for, 180–81, 185
waterproofing
    asphalt emulsion for, 152
    of block structures, 161–62
    of concrete, 100
    drainage and, 152, 154–56
    materials for, 156–61
    methods of, 163–64
    problems with, 84
    of wood houses, 162–63
waterproofing materials, 156–57,
    158–61
    bentonite, 84, 160–61, 163
    Bentonize, 161

liquid-applied membranes,
    156–57
    disadvantages of, 157
manufacturers of, 259–62
modified bitumens, 157
montmorillonite clays, 160–61
preformed membranes, 157,
    160
    installation of, 160
Volclay, 161
waterproofing methods, 163–64
waterstops, for cast-in-place con-
    crete, 118–19
water table
    measuring for, 98
    soil saturation and, 84–85
wind conditions, atrium houses and,
    24
windows
    clerestory
        advantages of, 16
        as light source, 209
    condensation on, 52
    crank-type casement, 67
    double-glazed, average net so-
        lar gain and, **47**
    placement of, 207–8
    reducing heat loss through,
        50–53
wiring, 174–79. *See also* electrical
    systems
    concealed, 178–79
    installation problems with,
        175–77
    surface-mounted, 177–78
wood, preservative for, 145, 147–48
wood construction, 147–51
    advantages of, 149–51
    drawbacks of, 150
    plumbing for, 150
    preservative for, 145, 147–48
    roof, 149
    waterproofing of, 151, 162–63
    wiring for, 150

wood heat
    as auxiliary heating system,
        187–89
    installation of, 190–91
    side effect of, 188
wood stoves
    as auxiliary heating system,
        187–89
    stack installation of, 190–91

wood structures, waterproofing of,
    162–63

Z

Zomeworks Corporation, manufac-
    turer of Nightwall, 53